PSYCHIATRY
FOR
GENERAL
PRACTITIONERS

PSYCHIATRY FOR GENERAL PRACTITIONERS

Editors:

R.C. Jiloha

Director Professor & Head
Department of Psychiatry
G.B. Pant Hospital
&
Maulana Azad Medical College
New Delhi–110002

M.S. Bhatia

Professor & Head
Department of Psychiatry
University College of Medical Sciences
&
Guru Teg Bahadur Hospital
Dilshad Garden
Delhi–110095

PUBLISHING FOR ONE WORLD

NEW AGE INTERNATIONAL (P) LIMITED, PUBLISHERS

New Delhi • Bangalore • Chennai • Cochin • Guwahati • Hyderabad
Jalandhar • Kolkata • Lucknow • Mumbai • Ranchi

Visit us at **www.newagepublishers.com**

Published by New Age International (P) Ltd., Publishers
First Edition: 2010

Branches:

- 36, Malikarjuna Temple Street, Opp. ICWA, Basavanagudi, **Bangalore**. ✆ (080) 26677815
- 26, Damodaran Street, T. Nagar, **Chennai**. ✆ (044) 24353401
- CE-39/1016, Carrier Station Road, Near Paulson Park Hotel, Ernakulam South, **Cochin.** ✆ (0484) 4051303
- Hemsen Complex, Mohd. Shah Road, Paltan Bazar, Near Starline Hotel, **Guwahati**. ✆ (0361) 2543669
- No. 105, 1st Floor, Madhiray Kaveri Tower, 3-2-19, Azam Jahi Road, Nimboliadda, **Hyderabad**. ✆ (040) 24652456
- RDB Chambers (Formerly Lotus Cinema) 106A,1st Floor, S.N. Banerjee Road, **Kolkata**. ✆ (033) 22275247
- 16-A, Jopling Road, **Lucknow**. ✆ (0522) 2209578
- 142C, Victor House, Ground Floor, N.M. Joshi Marg, Lower Parel, **Mumbai**. ✆ (022) 24927869
- 22, Golden House, Daryaganj, **New Delhi**. ✆ (011) 23262370, 23262368

ISBN: 978-81-224-2732-5

Rs. 495.00

C-09-06-3702

Printed in India at Glorious Printers, Delhi.
Typeset at In-house

PUBLISHING FOR ONE WORLD
NEW AGE INTERNATIONAL (P) LIMITED, PUBLISHERS
4835/24, Ansari Road, Daryaganj, New Delhi-110002
Visit us at **www.newagepublishers.com**

Preface

Psychiatry, being an allied branch of Medicine, is undergoing rapid changes. There have been many new advances in the causation, symptomatology, classification, diagnosis and management of mental disorders. The textbooks on psychiatric practice as applicable to Indian setting are few. A major portion of these textbooks is devoted to the psychiatric practice in western countries and is not practically oriented. Most of these textbooks do not contain the important aspects of psychiatry required by a General Practitioner.

The present book, "**Psychiatry for General Practitioners**" is an attempt to rectify most of these lacunae. A sincere effort has been made to make the book simple, easy, comprehensive and practically oriented. It also includes important common mental health problems faced by a General Practitioner in day today practice.

We record our sense of indebtedness and gratitude to the contributors and general practitioners for their constant inspiration and useful suggestions.

We hope that this book will be successful in fulfilling its aims. All suggestions are welcome and will be duly acknowledged.

Dr. R.C. Jiloha
Dr. M.S. Bhatia

List of Contributors

Chapter	*Contributors*
1. Introduction	**Dr. R.C. Jiloha,** Director Professor and Head, Department of Psychiatry, G.B. Pant Hospital and Maulana Azad Medical College, New Delhi-110002.
2. An Overview of Psychiatry	**Dr. M.S. Bhatia,** Prof. & Head, Department of Psychiatry, G.T.B. Hospital, Dilshad Garden, Delhi-110095.
3. Psychiatric Symptomatology, Interview and Examination	**Dr. M.S. Bhatia,** Prof. & Head, Department of Psychiatry, G.T.B. Hospital, Dilshad Garden, Delhi-110095.
4. Psychoses: Schizophrenia, Brief Psychotic Disorder and Delusional Disorder	**Dr. Smita N. Deshpande,** Senior Psychiatrist and Head, Depart-ment of Psychiatry, Dr. Ram Manohar Lohia Hospital, New Delhi-110001.
5. Mania and Bipolar Affective Disorder	**Dr. Rajesh Sagar,** Associate Professor, **Dr. Nitin Shukla,** Research Officer, Department of Psychiatry, All India Institute of Medical Sciences, New Delhi-110029.
6. Depression in General Practice	**Dr. Rakesh K. Chadda,** Professor of Psychiatry, All India Institute of Medical Sciences, New Delhi-110029.
7. Psychoactive Substance Abuse	**Dr. R.C. Jiloha,** Director Professor & Head, Department of Psychiatry, G.B. Pant Hospital and Maulana Azad Medical College, New Delhi-110002.
8. Anxiety Disorders	**Dr. Reshma,** Sr. C.M.O. Department of Psychiatry, G.B. Pant Hospital, New Delhi-110002.
9. Somatoform Disorders	**Dr. M.S. Bhatia,** Professor and Head, **Dr. Ravi Gupta,** Senior Resident, Department of Psychiatry, University College of Medical Sciences and Guru Teg Bahadur Hospital, Dilshad Garden, Delhi-110095.
10. Headache	**Dr. Ravi Gupta,** Senior Resident, **Dr. M.S. Bhatia,** Professor and Head, Department of Psychiatry, University

	College of Medical Sciences and Guru Teg Bahadur Hospital, Dilshad Garden, Delhi-110095.
11. Problems of Sleep	**Dr. Ravi Gupta,** Senior Resident, Department of Psychiatry, University College of Medical Sciences and Guru Teg Bahadur Hospital, Dilshad Garden, Delhi-110095.
12. Stress and its Management	**Dr. Shruti Srivastava,** Lecturer, University College of Medical Sciences and Guru Teg Bahadur Hospital, Dilshad Garden, Delhi-110095.
13. Psychosexual Disorders	**Dr. M.S. Bhatia,** Professor and Head, **Dr. Ravi Gupta,** Senior Resident, Department of Psychiatry, University College of Medical Sciences and Guru Teg Bahadur Hospital, Dilshad Garden, Delhi-110095.
14. Common Childhood and Adolescent Disorders	**Dr. Jitendra Nagpal,** Consultant Psychiatrist, VIMHANS, New Delhi.
15. Disorders Related to Women	**Dr. M.S. Bhatia,** Professor and Head, **Dr. Shruti Srivastava,** Lecturer, Department of Psychiatry, University College of Medical Sciences and Guru Teg Bahadur Hospital, Dilshad Garden, Delhi-110095.
16. Geriatric Psychiatry	**Dr. M.S. Bhatia,** Professor and Head, Department of Psychiatry, University College of Medical Sciences and Guru Teg Bahadur Hospital, Dilshad Garden, Delhi-110095.
17. Emergencies in Psychiatry	**Dr. Rajesh Rastogi,** Senior Psychiatrist, Safdarjung Hospital and V.M. Medical College, New Delhi-110029.
18. Culture Bound Syndromes in India	**Dr. Vishal Chhabra,** Senior Resident, Department of Psychiatry, University College of Medical Sciences and Guru Teg Bahadur Hospital, Dilshad Garden, Delhi-110095.
19. Legal and Ethical Issues in Psychiatry	**Dr. R.C. Jiloha,** Director Professor and Head, Department of Psychiatry, G.B. Pant Hospital and Maulana Azad Medical College, New Delhi-110002.
20. Psychopharmacology	**Dr. M.S. Bhatia,** Professor and Head, Department of Psychiatry, University College of Medical Sciences and Guru Teg Bahadur Hospital, Dilshad Garden, Delhi-110095.
21. Electroconvulsive Therapy	**Dr. M.S. Bhatia,** Professor and Head, Department of Psychiatry, University College of Medical Sciences and Guru Teg Bahadur Hospital, Dilshad Garden, Delhi-110095.
22. Psychological Methods of Treatment	**Dr. M.S. Bhatia,** Professor and Head, Department of Psychiatry, University College of Medical Sciences and Guru Teg Bahadur Hospital, Dilshad Garden, Delhi-110095.

Contents

	Preface	*v*
	List of Contributors	*vii*
1.	Intrduction	1–2
2.	An Overview of Psychiatry	3–6
3.	Psychiatric Symptomatology, Interview and Examination	7–28
4.	Psychoses : Schizophrenia, Brief Psychotic Disorder and Delusional Disorder	29–39
5.	Mania and Bipolar Affective Disorder	40–52
6.	Depression in General Practice	53–60
7.	Psychoactive Substance use Disorders	61–75
8.	Anxiety Disorders	76–80
9.	Somatoform Disorders	81–85
10.	Headache in General Practice: What you must know?	86–90
11.	Problems of Sleep	91–94
12.	Stress and its Management	95–100
13.	Psychosexual Disorders	101–114
14.	Common Childhood and Adolescent Disorders	115–127
15.	Disorder Related to Women	128–135

16. Geriatric Psychiatry 136–145

17. Emergencies in Psychiatry 146–152

18. Cultural Bound Syndromes in India 153–157

19. Legal and Ethical Issues in Psychiatry 158–164

20. Psychopharmacology 165–185

21. Electroconvulsive Therapy 186–190

22. Psychological Methods of Treatment 191–205

1

Introduction

Organisation of mental health services in the country has remained a subject matter of discussion in various workshops and seminars. The essential focus of these discussions has been the enormity of the mental health problems and the available technical know how in the country. The discrepancy between the magnitude of psychiatric problems in the general population and the number of psychiatrists available is quite evident. With the population of more than 100 crores, there are only 3000 psychiatrists in the country.

Situation is not any better in the city of Delhi. Considering the magnitude of psychiatric disorders, the services provided by a handful of government run hospitals and a few psychiatrists in private practice, are not adequate. A recent Need Assessment Survey (NAS) conducted by Delhi Mental Health Authority (DMHA) reveals that:

"Delhi has an estimated population of about 1,40,00,000. As per the WHO Report of 2000, about 25% of the general population suffers from psychiatric illnesses. Out of these about 1% suffer from schizophrenia and other psychotic disorders which are considered to be serious ailments while others suffer from depression, anxiety, adjustment disorders, substance abuse and other related disorders. Thus, approximately 14 lac persons in Delhi are in need of psychiatric help and out of these 1,40,000 are suffering from severe mental disorders. With the recent exodus of psychiatrists, Delhi is left with around 140 psychiatrists to take care of these patients. There are five Government Hospitals providing psychiatric services (including one psychiatric hospital) and 10 licenced psychiatric nursing homes to look after their indoor needs. There are only 452 psychiatric beds for Delhi population.

Situation is similar to other parts of the country. The National Mental Health Programme which came in 1982, strives to bring mental health services to each and every needy person. It could only be possible if mental health services are integrated with the general health services and with the community participation in the delivery of these services. The fact that there are only 1 or 2 psychiatrists per million populations in India there is a need for involving primary care doctors in the identification and management of common mental health problems.

As observed earlier, the training in psychiatry during undergraduate medical education is inadequate to identify and treat the mental illnesses.

Training of general practitioners in psychiatry under the National Mental Health Programme is one step towards effective delivery of mental health services to the general population. The department of psychiatry G.B. Pant Hospital has been identified by the Government of India as one of the centres to train the general practitioners under this programme. This programme was conducted for two years to train 800 general practitioners. We have drawn teaching faculty from various medical institutes of the city including All India Institute of Medical Sciences (AIIMS), University College of Medical Sciences (U.C.M.S.), Ram Manohar Lohia Hospital (RML), Safdarjung Hospital and VIMHANS. We have also invited senior and experienced psychiatrists in private sector to participate in this programme.

We have selected the topics of the book keeping in mind their utility in day to day practice, related to common problems encountered in the clinics such as depression, anxiety disorders, schizophrenia, mood disorders, drugs and alcohol problems, childhood behavioural disorders and others. We hope this book will disseminate the desired knowledge and confidence among the general practitioners in handing mental health problems and achieving the goal of delivering mental health services to each and every patient in need of it.

2

An Overview of Psychiatry

1. **Psychiatry:** The medical speciality concerned with the study, diagnosis, treatment and prevention of mental abnormalities and disorders. The word ***Psychiatry*** is derived from ***'psyche'***, the Greek word for ***soul or mind***, and ***'iatros'***, which is Greek for ***healer***. In Greek mythology, Psyche was a mortal woman made immortal by Zeus. The different branches in Psychiatry are:-
   (*a*) ***Child Psychiatry*****:** The science of healing or curing disorders of the psyche in children (i.e., those below 12 years of age). So is the psychiatry concerned with Adolescents—Adolescent Psychiatry.
   (*b*) ***Geriatric Psychiatry*****:** The branch of psychiatry that deals with disorders of old age; it aims to maintain old persons independently in the community as long as possible and to provide long-term care when needed.
   (*c*) ***Community Psychiatry*****:** The branch of psychiatry concerned with the provision and delivery of a coordinated program of mental health care to a specified population.
   (*d*) ***Forensic Psychiatry*** *(Legal Psychiatry)***:** Psychiatry in its legal aspects, including criminology, penology, commitment of the mentally ill, the psychiatric role in compensation cases, the problems of releasing information to the court, of expert testimony.
   (*e*) ***Social Psychiatry*****:** In Psychiatry, the stress laid on the environmental influences and the impact of the social group on the individual. The emphasis is on aetiology, purposes of treatment and prevention.
   (*f*) ***Cultural Psychiatry*** *(Comparative Psychiatry)***:** The branch of psychiatry concerned with the influence of the culture on the mental health of members of that culture. When the focus is on different cultures, the term transcultural psychiatry is used.
2. **Psychology:** The science that deals with the mind and mental processes—consciousness, sensation, ideation, memory etc.

3. **Psychodynamics:** The current usage of the term focuses on intrapsychic processes (rather than interpersonal relationships) and on the role of the unconscious motivation in human behaviour.
4. **Psychoanalysis:** A procedure devised by Sigmund Freud, for investigating mental processes by means of free association, dream interpretation, and interpretation of resistance and transference manifestations. A theory of psychology developed by Sigmund Freud out of his clinical experience with hysterical patients. A form of treatment developed by Sigmund Freud that utilises for psychoanalytic procedure and is based on psychoanalytic psychology.
5. **Psyche:** (Greek word meaning: 'The Soul') The mind.
6. **Mind:** It is the functional capacity of brain (brain is an anatomical structure.) e.g., Intelligence, memory. (It is divided into 3 components—Cognition (Intellect), Conation (psychomotor activity) and Affect (emotional part).
7. **Personality:** The characteristic way in which a person thinks, feels and behaves; the ingrained pattern of behaviour that each person evolves, both consciously and unconsciously as the style of life or way of being in adapting to the environment.
8. **Mental Health:** Psychological well-being or adequate adjustment, particularly as such adjustment conforms to the community accepted standards of behaviour.

Important characteristics of mental health are

— reasonable independence
— self-reliance
— self direction
— ability to do a job
— ability to take responsibility and make needed efforts
— reliability
— persistence
— ability to find recreation, as in hobbies
— satisfaction with sexual identity
— ability to work under authority, rules and difficulties
— a sense of competition, collaboration, compromise, satisfaction and security
— ability to show friendliness and love
— ability to give and take
— tolerance of others and of frustrations and emotions
— ability to contribute
— a sense of humour
— a devotion beyond oneself
— ability to get along with others

— cooperation
— optimism
— ability to function in both dependent and independent roles.

Psychiatrist: A medical graduate who has successfully undergone a postgraduate training course (of 2 to 3 years) in psychiatry.

Clinical Psychologist: A graduate in psychology who has successfully undergone a postgraduate training course (of 2 years) in clinical psychology.

Psychotherapist: A person with special training in psychotherapy. (Medical graduation is not a must).

Psychiatric Social Worker: A graduate in sociology who has successfully undergone a postgraduate training course of 2 years in social case work. This consists of knowledge and experiences in investigations of the social and cultural milieu of the patients and methods of correcting it whenever it is pathogenic.

Psychiatric Nurse: A nurse who has received special training in the care and management of psychiatric patients.

Occupational Therapist: A graduate who is trained in observing and treating the patients through crafts and recreational activities.

A mentally healthy person, while free of gross symptoms, and usually feeling well, is not always happy. The healthy adult may at times have some minor psychiatric symptoms. There are *clinical implications* of the concept of mental health i.e.,

- **Route examination of patients:**

  Evaluation of patients who are apparently not mentally ill but who wish professional help with personal problems.

- **Selection of treatment goals for psychiatric patients:**

  Vocational screening e.g. in Armed forces etc.

Community mental health activities (e.g. in formulating "Mental Health Act", Mental Health Programme etc.)

The problems not attributable to a mental disorder i.e., diagnosis which may be used for persons without mental disease, are

— marital problems
— other interpersonal problems
— phase of life problems
— other specified family circumstances
— academic problems
— noncompliance with medical treatment.

Others—uncomplicated bereavement, parent-child problems, anti-social behaviour, borderline intellectual functioning (usually I.Q. 70–80), malingering etc.

Magnitude of Problem in India

- Average prevalence of severe mental disorders is at least 18–20/1000 population; about 3–5 times that number suffer from other forms of distressing and socio-economically incapacitating emotional disorders. (The average prevalence of severe mental disorder is 2 per cent).
- 15–20% who visit general health services (such as a Medical OPD or a Private practitioner or a primary health centre) have emotional problems appearing as physical symptoms.
- Average number of new cases of serious mental disorders (Incidence) is about 35 per lac population.
- About 1–2% children suffer from learning and behaviour problems. Mental retardation estimated at 0.5–1.0% of all children.
- Among elderly (above 60 years of age) prevalence rate of mental morbidity is about 80–90/1000 population of aged (i.e., about 4 million severely mentally ill). This is in comparison to U.K. where the rate is as high as 260–265/1000. Geriatric depression is most frequent with a prevalence rate of 60/1000.
- Drug abuse surveys have reported the prevalence rate ranging from about 2–40% (Alcohol, tobacco, cannabis and opium are common).
- The common psychiatric illnesses encountered in a General Hospital Psychiatric Clinic are — Neuroses (Depressive neurosis followed by anxiety neurosis), Psychosomatic disorders (e.g. Peptic ulcer, Hypertension, Tension, Headaches etc.), Functional Psychoses (MDP depression, mania and schizophrenia) and organic psychoses (usually delirium). The other disorders such as Adjustment disorders, Psychosexual disorders are also not uncommon.
- In a Child Guidance Clinic, the common mental illnesses include mental retardation, emotional and behavioural (conduct) problems, enuresis, hyperkinetic syndrome etc. whereas in a Geriatric Clinic the common disorders are depression, dementia, paranoid disorders etc.
- In psychosexual clinics in India, the common problems encountered include 'Dhat syndrome', Premature ejaculation, Erectile impotence etc. whereas in de-addiction clinics, the patients who commonly come for treatment include Opiate dependence, Alcohol dependence, Polydrug abuse, Cannabis dependence etc.
- *Manpower.* In India, we have about 2500–3000 qualified psychiatrist, 600–700 psychologists, 300–400 psychiatric social workers and 700–800 psychiatric nurses.

3

Psychiatric Symptomatology, Interview and Examination

HISTORY TAKING

— The patient and the attendants must be helped to feel comfortable enough to give a detailed account of psychiatric disorder. A summary of history is given in Table 1.

Table 1: Summary of Psychiatric History

I. Identification of the patient	* Name, father's name, age, sex, literacy, marital status, religion, occupation, address, identification marks, photo, where seen (OPD/Ward), referral/direct.
II. Identification and reliability of informant	* Identification of patient, reliability (i.e., ability, to report, relationship, familiarity, length of stay with patient, attitude towards patient, history of physical/mental illness/drug abuse, reliability of information, intention for treatment)
III. Chief complaints or Reasons for referral and their duration	* According to patient/informant, duration, onset, course, predisposing precipitating factors
IV. History of present illness	
* Main problems (volunteered by patient and informants)	* Nature of problem (Psychiatric and Physical)
	* Date of onset
	* Duration

Contd...

		* Severity * Details as enlisted above.
	* Associated problems	* Also other disturbances (e.g. sleep, appetite, weight etc.) not narrated by the patient. * Also history of substance abuse.
	* Chronological development of problems	* How they developed? * When were they at their worst? * Any changes since onset? * And the factors or events responsible. * The factors increasing or decreasing severity.
	* Effects of the problem	* On activities such as — related personal hygiene — work — domestic tasks — leisure activities — social activities — family and other relationships — sexual activities.
	* Treatment taken so far	* Nature and effects of psychological or physical treatment given * Dose and duration of any psychotropic medication * Compliance * Time loss between onset of illness and treatment and reasons.
	* Patient's understanding of the problem	* Attitudes to understanding of problems.
	* Resources and strength	* Patient's family's and other helping resources and strength.
	* Legal issues	* Arrests, imprisonments, suspension from job etc.
	* Secondary gain	* Any gain, compensation or psychological
V.	**Past History**	
	* Psychiatric history	* Dates, main complaints or diagnosis, treatment. * Any complications e.g. suicidal attempts.

Contd...

* Completeness of recovery and socialisation/ personal care.
* History of drug abuse (types, duration, intoxication/withdrawal symptoms and treatment taken).

* Medical history
    * Chronic medical illnesses (e.g. diabetes mellitus etc.) and details of medication.
* Legal history
    * Any arrests, imprisonments, divorce, lawsuits etc.

VI. Family History

* **Types of family (Joint/Nuclear/Extended)**
* Parental history (Ages or age of death, health, mental/physical), occupation, social position, personality and relationship with patient)
* Family dynamics (history of mental illness)
* Relationship among family members; family events (initiating or exacerbating illness).

VII. Personal History

* **Birth**
* Early development
* Educational history
* Occupational history
* Menstrual history
* Sexual history
* Marital history
* Religious history
* Legal record

VIII. Premorbid Personality

* **Personality traits**
* Habits, hobbies, interests
* Beliefs, attitudes
* Social relationships
* Coping resources
* Alcohol drug abuse

IX. Checklist of Information Obtained

* **Any criminal record**
* Is history reliable?
* Is it complete?
* Any need to contact other informants?
* Areas needing mental status examination.

MENTAL STATUS EXAMINATION

I. APPEARANCE AND BEHAVIOUR

(a) Attitude

Describe the manner in which the patient relates to the examiner i.e.,

— cooperative
— trustful
— attentive
— seductive
— contentious
— ingratiating
— defensive
— friendly (Frank)
— purposefulness
— interested
— hostile
— playful
— evasive
— guarded

Is it possible to emphathise? (“***Empathy***” is defined as the ability to identify with the patient in order to recognise and identify the mental state). Schizophrenic patients may be difficult to be empathised.

Level of rapport (“***Rapport***” is a conscious feeling of accord, sympathy, trust and mutual responsiveness between one person and another).

(b) General Appearance and Grooming

— dressed with neatness
— clothes appropriate to season
— clothes clean and good
— hair/nails
— concern about appearance

(c) Facial Expression

— Is it ***appropriate or not?*** (Schizophrenic patients may have inappropriate or incongruous expression).
— Does it change with subject or not?
— ***Look***—attentive, apathetic (e.g. in chronic schizophrenic), indifferent (e.g. in severely depressive).
— ***Expression***—elation, fears, anger, sad, blank.
— ***Eye to Eye contact***—avoids gaze, excessive scanning.

(d) Posture

— Relaxed, Guarded, Sitting at the edge of the chair or in a picture for prolonged period (e.g. in schizophrenic patients).

(e) Gait and Carriage

— ***Type of Gait***—Normal, Brisk, Slow (e.g. in depressed patients), Desultory (e.g. in schizophrenic patients), Dilatory (e.g. in manic patients), Unsteady (e.g. in patients with organic brain disorders).

(f) Body Build

— ***Asthenic*** (Leptosomatic or ectomorph) i.e., Persons with narrow in length with narrow, shallow thorax with narrow subcostal angle are believed to be more prone to schizophrenia.

— ***Pyknic type*** (Endomorphic) i.e., Persons with large body cavities, relatively short limbs and large subcostal angle with rounded head and short, fat neck are believed to be more prone to manic depressive psychosis.

— ***Athletic type*** (Mesomorphic) i.e., Persons with wide shoulders and narrow hips and well developed bones and muscles are believed to be more prone to drug dependence etc.

(g) Psychomotor Activity

— ***Appropriateness***

— ***Quantity*: Normal, Increased (e.g. in mania, agitation), Decreased (e.g. in severe depression).

— ***Quality*:** Facial movements e.g. in oral dyskinesia, tremors in hands or body.

— ***Mannerisms*** i.e., odd, repetitive movements, may be a part of a goal directed activity (e.g. in normal persons, maniacs).

— ***Stereotypies*** i.e., Motor or verbal repetition without any discernible goal e.g. in schizophrenics.

— ***Automatic obedience*** (a type of catatonic behaviour when verbal instructions are overridden by tactile or visual stimuli e.g. the patient shakes hands with the examiner contrary to the firm verbal instructions whenever the examiner's right hand is extended) e.g. in schizophrenic patients.

— ***Mitmachen*** (Despite instructions to the contrary, the patient will allow a body part to be put into any position without resistance to the light pressure).

— ***Echopraxia*** e.g. in Catatonic Schizophrenics (Automatic copying of the examiner's movements or postures).

— ***Echolalia*** (Automatic repetition of the examiner's utterances) e.g. in catatonic schizophrenic patients.

— ***Catalepsy*** (Prolonged sustaining of an awkward posture or position) e.g. schizophrenic patients. (c.f. cataplexy, a type of sleep disorder).

— ***Cerea flexibilitas or waxy flexibility.*** (If the examiners encounter plastic resistance like the bending of a wax rod when moving the patient's arm, which will then be maintained in an odd position) e.g. in Catatonic Schizophrenic.

— ***Cogwheel or Lead pipe rigidity*** (e.g. in parkinsonism).

— ***Compulsions***—are obsessional motor acts (e.g. in obsessive compulsive disorder).

(h) Voice and Speech

- ***Intensity***
- ***Pitch*:** Monotonous (e.g. retarded patients), Abnormal changes (e.g. in manic patients).
- ***Speech*:** Slow (e.g. in depressive), Rapid (e.g. manic patient).
- ***Ease of speech***
- ***Spontaneity***
- ***Productivity***
- ***Relevance***
- ***Manner***
- ***Deviation*:** *Neologism* (coining of new words or used words in an inappropriate way), Echolalia, *Clang association* (Speech in which sounds rather than meaningful conceptual relationships govern word choice) e.g. mania. *Verbigeration* (a manifestation of stereotypy consisting of morbid repetition or words, phrases or sentences also called cataphasia) e.g. in schizophrenia.
- ***Reaction time***
- ***Vocabulary and diction***

Table 2: Outline of Mental Status Examination

I.	**Attitudes, Appearance and Behaviour**		
	a. Attitudes	:	Type (cooperative, friendly, trustful, attentive, interested, seductive, evasive, defensive, guarded) Is it possible to empathise with patient? Level of rapport
	b. General appearance and grooming	:	Dressing, Personal hygiene
	c. Facial expression	:	Is it appropriate or not? Does it change with subject or not? Expression types. Eye to eye contact
	d. Posture		
	e. Gait and carriage		
	f. Body build		
	g. Psychomotor activity	:	Appropriateness, quantity, quality.
	h. Voice and speech	:	Intensity, quantity, quality, spontaneity, relevance, reaction time, vocabulary

Contd...

II. Mood and Affect	:	Appropriateness (Quality), Intensity (Quality), Type, Range, Stability, Relatedness
III. Perception	:	Illusions, hallucinations, depersonalisation, derealisation, others (deja vu, deja pense, deja entendu, jamais vu etc.)
IV. Thinking	:	Stream (tempo/continuity) Possession (obsession/phobias, overvalued ideas) Content (Primary or Secondary delusional disorder), Form
V. Judgement	:	Test judgement Social judgement
VI. Insight	:	Degree of insight Intellectual/emotional insight
VII. Sensorium and Cognition	:	Consciousness Orientation (Time, Place, Person) Attention, Concentration Memory (Remote, Recent, Immediate) Abstraction General intelligence

II. MOOD AND AFFECT

Mood

It is a sustained subjective feeling state, which can be described by qualities such as happiness, sadness, worry, anxiety, irritability, anger, detachment and indifference.

Affect

It refers to more transitory and immediate emotional expressions with described mood content as one component (i.e., pleasurable or unpleasurable feeling tone associated with an idea, image or a situation).

Aspects of Affect

— ***Appropriateness (Quality)*:** It refers to affective display to the content of speech and thought. Inappropriate affect is characteristic of schizophrenia (where it is also known as *incongruous affect*). Affect may be inappropriate and labile in pseudobulbar palsy.

— ***Intensity (Quantity)*:** Normal, increased as in vituperative, ineffective speech with dogmatic insistence regarding self-convictions e.g. in mania.

— ***Type*:** Elated (e.g. in mania), Sad (e.g. in depression), Fear (e.g. in phobia), Anger (e.g. in schizophrenia), Anxious (e.g. in anxiety neurosis), Irritable (e.g. in hypomania), *Blunting* (affects are diminished in intensity and constricted to a narrow neutral range) e.g. in schizophrenia. *Flat* (No affective response at all) e.g. in schizophrenia.

— ***Range*:** Constricted e.g. in depression. Expanded e.g. in mania.

— ***Stability*:** Stable, Liability e.g. in organic mental disorders.

III. PERCEPTION

(a) Illusions

These are misinterpretations of perceptions (e.g. shadows are mistaken for frightening figures). Illusion may occur in normal individuals or in organic mental disorders.

(b) Hallucinations

Perceptions occurring without external stimulation. Hallucinations may depend on type of sensory system affected e.g. auditory, visual, olfactory, gustatory. ***Functional hallucinations*:** Occur only when there is a concurrent real perception in the same sensory modality (e.g. hearing voices only when the water tap is on). ***Autoscopic hallucinations*:** A visual hallucination of patients themselves. ***Extracampine hallucinations*:** When occurring outside of a known sensory field (e.g. seeing objects through a solid wall). ***Kinaesthetic hallucinations*:** Feeling movement when none occurs e.g. out of body experiences. ***Hypnagogic hallucinations*:** which occur when falling asleep. ***Hypnopompic hallucinations*:** which occur when awakening.

(c) Depersonalisation and Derealisation

These are alternations in the perception of one's reality.

— ***Depersonalisation*:** The patient feels detached and views himself or herself as strange and unreal. (It is an "as if" phenomenon and patient is not fully convinced).

— ***Derealisation*:** It involves a similar alteration in the sense of reality of the outside world. (Familiar objects or places may seem altered in size and shape).

(d) Other Abnormal Perceptions

— ***Deja vu*:** Feeling of familiarity with unfamiliar things.

— ***Deja pense*:** A patient's feeling, verging on certainty that he has already thought of the matter.

— ***Deja entendu*:** The feeling that one had at some prior time heard or perceived what one is hearing in the present.

— ***Jamais vu*:** An erroneous feeling or conviction that one has never seen anything like that before (i.e., Feeling of unfamiliarity).

IV. THINKING

(a) Stream

It includes

(*i*) ***Disorders of Tempo***

— ***Flight of ideas*** (Rapid speech with quick changes of ideas) that may be associated by chance factors such as by the sound of the words but which can usually be understood. e.g. (See the king is standing, king, king, sing, sing, bird on the wing, wing wing). e.g. in mania.

— ***Inhibition or retardation***

— ***Circumstantiality***: Thinking proceeding slowly with many unnecessary trivial details but finally the goal is reached e.g. in mania, organic mental disorders, schizophrenia.

— ***Tangentiality***: It differs from circumstantiality that the final goal is not reached and the patient loses track of the original question.

— ***Incoherence***: Marked degree of loosening of associations in which the patient shifts ideas from one to another with no logical connection, accompanied by a lack of awareness on the part of the patient that ideas are not connected. It is seen in schizophrenia.

(*ii*) ***Disorders of Continuity***

— ***Perseveration***: Mental operations tend to persist beyond the point at which they are relevant. e.g. repetition of the same words or phrases over and over again despite the interviewer's direction to stop.

— ***Blocking***: Occurs when the thinking process stops altogether. It occurs in schizophrenia and anxiety states. The patients may or may not start with the same topic again.

— ***Echolalia***: Repetition of the interviewer's words, like a parrot.

(b) Possession and Control

— ***Obsessions***: Persistent occurrence of ideas, thoughts, images, impulses or phobias.

— ***Phobias***: Persistent, excessive, irrational fear about a real or an imaginary object, place or a situation.

— ***Thought alienation***: The patient has the experience that his thoughts are under the control of an outside agency or that others are participating in his thinking. It may be insertion, withdrawal or broadcasting.

— ***Suicidal/homicidal thoughts.***

(c) Content

(*i*) ***Primary Delusions***

Fixed unshakable false beliefs, which are against one's sociocultural and educational background, and they cannot be explained on the basis of reality. The patients lack insight into it. Primary delusions can be:

— ***Delusional mood:*** Patient thinks that something is going on around him which concerns him but he does not know what it is.

— ***Delusional perception:*** Attribution of a new meaning usually in the sense of self-reference, to a normal perceived object. It cannot be understood from one's affective states or previous attitudes, e.g. patient hears the stairs creak and knows that this is a detective spying on him.

— ***Sudden delusional ideas:*** A sudden revelation or well-formed ideas appear in the thinking e.g. a patient says that he of royal descent because he remembers when he was taken to a military parade as a little boy, the king saluted him.

(*ii*) ***Secondary Delusions***

They arise from some other morbid experience, e.g. delusions. For example, the patient unconsciously thinks 'I love him—I do not love him—I hate him—he hates me.'

(*iii*) ***Content of Delusions***

Delusions of: persecution, self-reference, grandiosity, ill health or somatic function, guilt, nihilism (the patient denies the existence of his body, his mind or the world around), poverty, love or erotomania.

(d) Form of Thinking

(*i*) ***Given by Cameron***

— ***Asyndesis:*** Lack of adequate connections between successive thoughts.

— ***Metonyms:*** Imprecise expressions or use of substitute term or phrase instead of more exact one (e.g. for a pen-writing stick).

— ***Interpenetration of themes:*** The patient's schizophrenic's speech contains elements which belong to the task in hand interspersed with a stream of phantasy which he cannot understand.

— ***Overinclusion:*** Inability to maintain the boundaries of the problem and to restrict operations within their correct limits. The patient shifts from one hypothesis to another.

(*ii*) ***Given by Goldstein***

In schizophrenia and organic mental disorders, there is a loss of abstract form i.e., thinking becomes *concrete* (patient is unable to free himself from the superficial concrete aspects of thinking).

(*iii*) ***Given by Schneider*** (Mnemonic *'FODDS'*)

— ***Substitution*** (A major thought is replaced by a subsidiary one).

— ***Omission*** (senseless omission of a thought or part of it).

— ***Fusion*** (Heterogeneous elements of thoughts are interwoven).

— ***Drivelling*** (Disorganised intermixture of constituent parts of one complete thought).

— ***Desultory thinking*** (speech in grammatically and syntactically correct but sudden ideas force their way from time to time).

V. JUDGEMENT

It is the capacity to draw direct conclusions from the material acquired by experience. It is impaired in psychoses.

- — ***Test judgement*** (Patient's prediction of what he or she would do in imaginary situations).
- — ***Social judgement*** (Subtle manifestations of behaviour that are harmful to the patient and contrary to acceptable behaviour in the culture).

VI. INSIGHT

It refers to subjective awareness of the pathological nature of psychiatric symptoms and behavioural disturbances. Lack of insight is characteristic of psychoses.

Clinical Rating of Insight

Insight is rated on a 6-point scale from one to six.

1. Complete denial of illness.
2. Slight awareness of being sick and needing help, but denying it at the same time.
3. Awareness of being sick, but it is attributed to external or physical factors.
4. Awareness of being sick, due to something unknown in self.
5. ***Intellectual insight*:** Awareness of being ill and that the symptoms/failures in social adjustment are due to own particular irrational feelings/thoughts; yet does not apply this knowledge to the current/future experiences.
6. ***True emotional insight*:** It is different from intellectual insight in that the awareness leads to significant basic changes in the future behaviour and personality.

VII. SENSORIUM AND COGNITION

(a) Consciousness

Awareness of facts and the content to mental phenomenon and degree of reactivity of the environment. Consciousness may be clouded e.g. in organic mental disorder.

(b) Orientation

Awareness about time, place and person. When it is impaired, it is usually in the order of time, then place and then person and when it is regained it is in the reverse order.

(c) Attention and Concentration

- — ***Active attention (concentration)*:** The amount of effort the patient exerts to solve a problem. It is tested by asking the patient to solve certain problems. (e.g. keep on subtracting seven from 100 or 4 times 5 or months of the year backwards).
- — ***Passive attention*:** The attention, which the environment draws, and the patient pays very little effort e.g. a shop on fire, an accident.

(d) Memory

The ability, process or act of remembering of recalling and especially the ability to reproduce what has been learned or experienced. The memory can be:

— ***Remote memory*****:** The ability to recall information what was experienced in the distant past.

— ***Recent past memory*****:** The past few months.

— ***Recent memory*****:** The past few days, recall of what was done yesterday, the day before, what was eaten for breakfast, lunch, dinner etc.

— ***Immediate retention and recall*****:** The ability to register information. Ability to repeat 5–6 figures after examiner dictates them—first forward and then backward, then after a few minutes interruption.

The memory may be impaired in organic mental disorders (dementia) or amnesia (organic and psychogenic).

(e) Abstraction (Abstract Thinking)

It is determined by asking the meaning of common (prevalent in a culture) idioms, proverbs and similarities and differences between objects in the same class, e.g. similarities and differences between "ball and orange" "fly and aeroplane" etc.

(f) General Intelligence

It can be gauged by patient's vocabularies, complexity of concepts they use and progressively more difficult questions about current events.

(g) Attitudes and Beliefs

It is important to note patient's attitudes and beliefs towards

— the illness

— the consequences of and limitations imposed by the illness.

— any help offered.

PHYSICAL EXAMINATION

The physical examination of the psychiatric patient is no less important than that of any sick person.

(a) General Physical Examination

* General examination (Skin, teeth, special senses).
* Basic parameters (Pulse rate, Blood pressure, Respiratory rate, Temperature and Fundus occuli).
* Look for pallor, icterus, oedema, lymphadenopathy.

Comments: Some common illnesses e.g. anaemia, dehydration, pyrexia etc. may present with symptoms mimicking psychiatric illness (e.g. anxiety attacks, phobias etc.)

(b) Systemic Examination: It includes examination of various systems, e.g.

i. ***Cardiovascular system*:** apex beat, regularity, heart sounds, murmurs.

N.B. 1. Some disorders such as Paraxysmal atrial tachycardia, Mitral valve prolapse may present as panic attacks or syncope.

2. Some treatments have to be avoided in certain CVS disorders (e.g. Electro-convulsive therapy in recent MI, tricyclic antidepressants in arrhythmias etc.).

ii. ***Respiratory system*:** Chest expansion on both sides, percussion, Adventitious sounds.

N.B. Some respiratory disorder e.g. asthma, bronchiolitis may present with symptoms mimicking anxiety and other disorders.

iii. ***Abdomen*:** Tenderness, bowel sounds, organomegaly, hernias etc.

N.B. Psychotropic drugs interfere with GIT motility and may produce gastritis, hepatotoxicity, constipation (leading to precipitation of piles, hernias etc.).

iv. ***CNS*:** It consists of

(a) *General observations.* Position of body, head, extremities; shape, tenderness, percussion of head; tenderness and rigidity of neck. Look for gait abnormalities.

(b) *Cranial nerves.* Look for palsies, neuralgias.

(c) *Corpus striatum.* Muscular rigidity, tremors or involuntary movements, akinesia, change of emotional expression.

(d) Cerebellum, station, Romberg sign, gait, hypotonicity, nystagmus, dysarthria, ataxia (finger to nose or finger, past-pointing, adiadokokinesis).

(e) Spinal cord and body segmental representation.

- *Sensory system*: Pain, temperature, light touch, deep touch, vibration, tactile discrimination.
- *Motor system*: Range of movements, contratures, atrophy, strength of muscles, tremors etc.
- *Reflexes*: Superficial (abdominal, cremastric, Babinski, Oppenheim, Gordon), Deep (biceps, triceps, knee, ankle, radial, ankle clonus etc.).

v. ***Musculoskeletal*:** For example, Pain and swelling in joints, neck pain, backache, myalgias.

N.B. Some musculoskeletal disorders e.g. Cervical spondylosis may present with symptoms mimicking migraine, anxiety states, hysteria (fainting) etc.

DIAGNOSTIC FORMULATION

It consists of:

(*i*) Summary of patient's problems.

(*ii*) Salient features of genetic, constitutional, familiar and environmental influences.

(*iii*) Important findings (Positive and negative) on mental examination.

(*iv*) Provisional diagnosis and differential diagnosis.

TREATMENT PLAN

It should stress on:

(*i*) The problems needing urgent attention (e.g. excitement, stupor, suicidal ideation etc.)

(*ii*) The reasons for hospitalisation (if any).

(*iii*) Investigations or tests required.

(*iv*) Treatments e.g. Medication (injectable or oral), Physical treatment (Electroconvulsive therapy), Psychotherapy, behavioural modification, counselling of relatives etc. and their duration.

(*v*) Prognosis. Favourable and poor prognostic factors.

(*vi*) Others: Insist on continuous supervision of patient by the close relative; compliance with treatment maintenance of hygiene and avoidance of sharp instruments, rope or live electric wires.

GLOSSARY

Abstraction: The process whereby thoughts or ideas are generalised and dissociated from particular concrete instances or material objects. Concreteness in proverb interpretation suggests an impairment of abstraction, as in schizophrenia.

Affect: The subjective and immediate experience of emotion attached to ideas of mental representations of objects. Affect has outward manifestations that may be classified as restricted, blunted, flattened, appropriate, or inappropriate.

Affect, Abnormal: A general term describing morbid or unusual mood states of which the most common are depression, anxiety, elation, irritability and affective lability.

Affect, Blunted: A disturbance of affect manifested by a severe reduction in the intensity of externalised feeling tone. Observed in schizophrenia. It is one of that disorder's fundamental symptoms, as outlined by ***Eugen Bleuler.***

Affect, Flat: Absence or near absence of any signs of affective expression. This may occur in schizophrenia, dementia or psychopathic personality.

Affect, Inappropriate: Emotional tone that is out of harmony with the idea, thought or speech accompanying it.

Affect, Labile: Affective expression characterised by repetitious and abrupt shifts, most frequently seen in organic brain syndromes, early schizophrenia and some forms of personality disorders.

Affect, Restricted: Affective expression characterised by a reduction in its range and intensity.

Affect, Shallow: A state of morbid sufficiency of emotional response presenting as an indifference to external events and situations, occurring characteristically in schizophrenia of the hebephrenic type but also in organic cerebral disorders, mental retardation and personality disorders.

Aggression: Forceful physical, verbal or symbolic action. May be appropriate and self-protective, including healthy self-assertiveness or inappropriate as in hostile or destructive behaviour.

Agitation: Excessive motor activity, usually non-purposeful and associated with internal tension. Examples, inability to sit still, fidgeting, pacing, writhing of hands or pulling of clothes.

Agnosia: Inability to understand the importance of significance of sensory stimuli cannot be explained by a defect in sensory pathways or sensorium.

Agoraphobia: Fear of open places: as phobic disorder characterised by a fear of leaving one's home. It may present with or without panic attacks. It is the *commonest form* of phobia, seen in clinical practice. Psychological treatments may attempt either to reduce the symptoms of the phobia or to resolve the underlying anxiety.

Agraphia: Loss of impairment of a previously possessed ability to write; may follow parietal lobe damage.

Akathisia: A state of motor (or less often verbal) restlessness manifested by the compelling need to be in constant movement. It may be seen as an extrapyramidal side effect of *butyrophenone or phenothiazine* medication.

Akinesia: Lack of physical movement, as in the extreme immobility of catatonic schizophrenia.

Ambitendence: A psychomotor disturbance characterised by an ambivalence towards a voluntary action, leading to contradictory behaviour, most frequently seen in catatonic schizophrenia.

Amnesia: Pathologic loss of memory; a phenomenon in which an area of experience becomes inaccessible to conscious recall. It may be organic, emotional or of mixed origin and limited to a sharply circumscribed period of time. Two types are: *retrograde*: Loss of memory for events preceding the amnesia proper and the condition(s) presumed to be responsible for it; *anterograde*: Inability to form new memories following such condition(s).

Anxiety: Unpleasurable emotional estate associated with psychophysiological changes in response to an intrapsychic conflict, in contrast to fear, the danger or threat in anxiety is unreal.

Apathy: Want of feeling or affect or interest or emotional involvement in one's surroundings. It is observed in certain types of schizophrenia and depression.

Aphasia: A disturbance in language function due to organic brain disorder.

Apperception: Awareness of the meaning and significance of a particular sensory stimulus as modified by one's own experiences, knowledge, thoughts and emotions. See also perception.

Apraxia: Inability to perform a voluntary purposeful motor activity. The inability cannot be explained by paralysis or sensory impairment.

Ataxia: Lack of coordination, either physical or mental. In neurology, it refers to loss of muscular coordination. In psychiatry, the term *'intrapsychic ataxia'* refers to lack of coordination between feelings and thoughts; the disturbance is found in schizophrenia.

Attention: Concentration, the aspect of consciousness that relates to the amount of effort exerted in focussing on certain aspects of an experience, activity or task.

Attitude: A 'mental set' held by an individual which affects the ways that person responds to events and organises his cognitions.

Automatic obedience: The phenomenon of undue compliance with instruction, a feature of command automatism associated with catatonic syndromes and the hypnotic state.

Automatism: Automatic and apparently undirected non-purposive behaviour that is not consciously controlled. Seen in the psychomotor epilepsy.

Awareness: A subjective state of being *alert or conscious;* cognisant of information received from the immediate environment.

Aypnia: Insomnia; inability to sleep.

Catalepsy: Condition in which a person maintains the body position which he is placed. It is a symptom observed in severe cases of catatonic schizophrenia. It is also known as *waxy flexibility* and *cerea flexibilitas.* See also Command automatism.

Cataplexy: Temporary sudden loss of muscle tone, causing weakness and immobilisation. It can be precipitated by a variety of emotional states, and it is often followed by sleep.

Catathymia: A situation in which elements in the unconscious are sufficiently affected to produce changes in conscious functioning.

Cerea flexibilitas: The waxy flexibility often present in catatonic schizophrenia in which the patient's arm or leg remains in the position in which it is placed.

Circumstantiality: Disturbance in the associative thought and speech processes in which the patient digresses into unnecessary details and inappropriate thoughts before communicating the central idea. It is observed in schizophrenia, obsessional disturbances, and certain cases of dementia. See also Tangentiality.

Clang association: Association or speech directed by the sound of a word, rather than its meaning. Punning and rhyming may dominate the person's verbal behaviour. It is seen most frequently in schizophrenia or mania. Also known as *clanging.*

Cognition: Mental process of knowing and becoming aware. One of the ego functions. It is closely associated with judgement. Groups that study their own processes and dynamics use more cognition than the encounter groups, which emphasise emotions. It is also known as *thinking.*

Command automatism: Condition closely associated with catalepsy in which suggestions are followed automatically.

Compulsion: Uncontrollable, repetitive and unwanted *urge* to perform the act. It serves as a defense against unacceptable ideas and desires, and failure to perform the act leads of to overt anxiety. See also Obsession, Repetition compulsion.

Conation: That part of person's mental life concerned with his strivings, motivations, drives and wishes as expressed through his behaviour.

Concrete thinking: Thinking characterised by actual things and events and immediate experience, rather than by abstractions. Concrete thinking is seen in young children: in those who have lost or never developed the ability to generalise as in certain organic mental disorders; and in schizophrenics. See also Abstract thinking.

Confabulation: Unconscious filling the gaps in memory by imaging experiences or events that have no basis in fact. It is common in organic amnestic syndromes. Confabulation should be differentiated from lying. See also Paramnesia.

Conflict: A mental struggle that arises from the simultaneous operation of opposing impulses, drives, external (environmental) or internal demands. Termed *intrapsychic* when the conflict is between forces within the personality; *extrapsychic,* when it is between the self and the environment.

Confusion: A term usually employed to designate a state of impaired consciousness associated with acute or chronic cerebral organic disease. Clinically it is characterised by disorientation, slowness of mental processes with scanty association of ideas, apathy, lack of initiative, fatigue and poor attention.

Congruence: A general term used to refer to behaviour, attitudes or ideas which are in accord and not in conflict with other such behaviour, attitudes or ideas.

Conscience: The morally self-critical part of one's standards of behaviour, performance and value judgements. Commonly equated with the superego.

Consciousness: The awareness of one's own mental processes, or the state of having this awareness.

Constitution: A person's intrinsic psychological or physical endowment.

Conversion: *A defense mechanism,* operating unconsciously, by which intrapsychic conflicts that would otherwise give rise to anxiety are, instead, give symbolic external expression.

Coprolalia: The use of vulgar or obscene language.

Deja entendu: Illusion that what one is hearing one has heard previously.

Deja pense: A condition in which a thought never entertained before is incorrectly regarded as a repetition of a previous thought.

Deja vu: Illusion of visual recognition in which a new situation incorrectly regarded as a repetition of a previous experience. See also (Paramnesia).

Delirium: An acute, reversible organic mental disorder characterised by confusion and some impairment of consciousness.

Delusion: A false belief that is firmly held, despite objective and obvious contradictory proof or evidence and despite the fact that other members of the culture do not share the belief. Types of delusion include *Bizarre delusion.* False belief that is patiently absurd or fantastic. *Delusion of control.* Delusion that a person's thoughts, feelings, or actions are not his own but are being imposed on him by some external force, *Delusion of grandeur (grandiose delusion).* Exaggerated concept of one's importance, power, knowledge, or identity. *Delusion of jealousy (delusion of infidelity).* Delusion that one's lover is unfaithful. *Delusion of persecution.* Delusion that one is or will be without material possessions. *Delusion of reference.* Delusion that events, objects, or the behaviour of others have a particular and unusual meaning specifically for oneself. *Encapsulated delusion.* Delusion without significant effect on behaviour. *Fragmentary delusion.* Poorly elaborated delusion, often one of many with no apparent interconnection. *Nihilistic delusion (delusion of negation).* Depressive delusion that the world and everything related to it have ceased to exist. *Paranoid delusion.* Delusion of persecution grandiose delusion. *Religious delusion.* Delusion involving the Deity or theological themes. *Sexual delusion.* Delusion centering on sexual identity, appearances, practices, or ideas. *Somatic delusion.* Delusion pertaining to the functioning of one's body. *Systemised delusion.* A group of elaborate delusions related to a single event or theme.

Dementia: An organic mental disorder characterised by general impairment in intellectual functioning. *Frequent components* of the clinical syndrome are failing memory, difficulty with calculations, distractibility, alterations in mood and affect, impairment in judgement and abstraction, reduced facility with language, and disturbance of orientation.

Depersonalisation: Sensation of unreality concerning oneself, parts of oneself, parts of oneself, or one's environment which occurs under extreme stress or fatigue. It is seen in schizophrenia, depersonalisation disorder, and schizotypal personality disorder. See also Ego boundaries.

Depression: A mental state characterised by feeling of sadness, loneliness, despair, low self-esteem, and self-approach. Accompanying signs include psychomotor retardation or at times agitation, withdrawal from interpersonal contact, and vegetative symptoms, such as insomnia and anorexia.

Derealisation: Sensation of changed reality or that one's surroundings have altered. It is usually seen in schizophrenics. See also Depersonalisation.

Dissociation: The splitting off of clusters of mental contents from conscious awareness, a mechanism central to hysterical conversion and dissociative disorders; the separation of an idea from its emotional significance and affect as seen in the inappropriate affect of schizophrenic patients.

Doctor-patient relationship: Human interchange that exists between the person who is sick and the person who is selected because of training and experience to heal.

Dystonia: *Extrapyramidal motor disturbance* consisting of slow, sustained contractions of the axial or appendicular musculature; one movement often predominates, leading to relatively sustained postural deviations. Acute dystonic reactions (facial grimacing, torticollis) are occasionally seen with the initiation of antipsychotic drug therapy.

Echolalia: Repetition of another person's words or phrase. Observed in certain cases of schizophrenia, particularly the catatonic types. The behaviour is considered by some authors to be an attempt by the patient to maintain a continuity of thought processes. See also Communication disorder, Gilles de la Tourette's disease.

Echopraxia: Repetition of another person's movements. It is observed in some cases of schizophrenia.

Emotion: The experience of subjective feelings which have positive or negative value for the individual.

Empathy: The intellectual and emotional awareness and understanding of another person's state of mind. It involves the *projection* of oneself into another person's frame of reference. It is important ability in a successful therapist or a helpful group member. See also Sympathy.

Extroversion: The state of one's energies being directed outside oneself. It is also spelled as extraversion.

Flight of ideas: A nearly continuous flow of accelerated speech with abrupt changes from topic to topic, usually based on understandable associations, distracting stimuli, or plays on words. When severe, the speech may be disorganised and incoherent. Flight of ideas is most frequently seen in *Manic episodes,* but may also be observed in some cases of *Organic Mental Disorders, Schizophrenia,* other *psychotic disorders,* and occasionally, acute reactions to stress.

Floccillation: Aimless plucking or picking, usually at bedclothes or clothing. It is common in senile psychosis and delirium.

Forgetting: Broadly speaking, theories of forgetting can be sorted into seven major approaches; decay theory (the idea that memory traces gradually decay overtime, unless strengthened by being retrieved); *interference theory; amnesia* brought about through physical causes; *motivated forgetting;* lack of appropriate cues for retrieval; lack of the relevant context of retrieval; and inadequate processing during storage.

Formal thought disorder: A disturbance in the form of thought as distinguished from the content of thought. The boundaries of the concept are not clear and there is no consensus as to which disturbances in speech or thoughts are included in the concept. For this reason, "formal thought disorder" is not used as a specific descriptive term in DSM-IV.

Formication: A tactile hallucination involving the sensation that tiny insects are crawling over the skin. It is most commonly encountered in *cocainism* and *delirium tremens.*

Hallucination: A false sensory perception occurring in the absence of any relevant external stimulation of the sensory modality involved. Examples include: *Auditory hallucination.* Hallucination of sound. *Gastatory hallucination.* Hallucination of taste. *Hypnagogic hallucination.* Hallucination occurring while awaking for sleep (ordinarily not considered pathological). *Kinaesthetic hallucination.* Hallucination of bodily movement. *Lilliputian hallucination.* Visual sensation that persons or objects are reduced in size, it is more properly regarded as an illusion (see also Micropsia). *Olfactory hallucination.* Hallucination involving smell. *Somatic hallucination.* Hallucination involving the perception of a physical experience localised within the body. *Tactile (haptic) hallucination.* Hallucination involving the sense of touch. *Visual hallucination.* Hallucination involving sight.

Idea: The memory of past perceptions. An idea depends upon an image in the same way as a perception depending upon a sensation.

Illusion: Perceptual misinterpretation of a real external stimulus.

Incoherence: Speech that, for the most part, is not understandable, owing to any of the following: a lack of logical or meaningful connection between words, phrases, or sentences, excessive use of incomplete sentences' excessive irrelevancies or abrupt changes in subject matter; idiosyncratic word usage; distorted grammar. Mildly ungrammatical construction or idiomatic usages characteristic of particular regional or ethnic backgrounds, lack of education, or low intelligence should not be considered coherence; and the term is generally not applied when there is evidence that the disturbance in speech is due to an aphasia. Incoherence may be seen in some *Organic Mental Disorders, Schizophrenia,* and other *psychotic* disorders.

Insight: Conscious recognition of one's own condition. In psychiatry, it more specifically refers to the conscious awareness and understanding of one's own psychodynamics and symptoms of maladaptive behaviour. It is highly important in effecting changes in the personality and behaviour of a person. *Intellectual insight* refers to knowledge of the reality of a situation without the ability to successfully use that knowledge to effect an adaptive change in behaviour. *Emotional insight* refers to deeper level of understanding or awareness that is more likely to lead to positive change in personality and behaviour.

Intelligence: The capacity for learning and the ability to recall, integrate constructively, and apply what one has learned; the capacity to understand and to think rationally.

Introvert: An individual inclined towards a solitary, reflective life-style.

Jamais vu: False feeling of *unfamiliarity* with a real situation that one has experienced; it is a paramnestic phenomenon. See also Paramnesia.

La belle indifference: An inappropriate attitude of calm or lack of concern about one's disability. It is seen in patients with conversion disorder. See also Hysterical neurosis. Literally "*beautiful indifference*."

Loosening of associations: A characteristic schizophrenic thinking or speech disturbance involving a disorder in the logical progression of thoughts, manifested as a failure to adequately verbally communicate. Unrelated and unconnected ideas shift from one subject to another.

Mental disorder: A psychiatric illness or disease. Are manifestations primarily behavioural or psychological? It is measured in terms of deviation from some normative concept.

Mental retardation: A condition of arrested or incomplete development of the mind which is especially characterised by subnormality of intelligence.

Mood: A pervasive and sustained emotion that in the extreme, markedly colours the person's perception of the world. *Mood is to affect as climate is to weather,* common examples of mood include depression, elation, anger and anxiety.

Mood-congruent psychotic features: A DSM-III term which refers to hallucinatory or delusional phenomena whose content consistently reflects the mood of a manic or depressed patient. See also Nihilism.

Mood-incongruent psychotic features: A DSM-II term which refers to hallucinatory or delusional phenomena whose content consistently reflect the mood of a manic or depressed patient.

Negativism: Verbal or nonverbal opposition or resistance to outside suggestions and advice. It is commonly seen in catatonic schizophrenia in which the patient resists any effort to be moved or does the opposite of what is asked. It may also occur in organic psychoses an mental retardation.

Neologism: New word or phrase, often seen in schizophrenia. Definitions restrict the use of the term to those new words or phrases whose derivation cannot be understood. However, the term "neologism" has also used to mean a word that has been incorrectly constructed but whose origins are nonetheless understandable, for example, "headshoe" to mean "hat." Those words are more properly referred to as word approximations. See also Metonymy, Paraphysis, Word approximation.

Obsession: Persistent and recurrent idea, thought, or impulse that cannot be eliminated from consciousness by logic or reasoning. Obsessions are involuntary and egodystonic. See also Compulsion.

Paraphasia: Type of abnormal speech in which one word is substituted for another, the irrelevant word generally resembles the required one in its morphology, meaning, or phonetic composition.

Perception: The conscious awareness of elements in the environment by the mental processing of sensory stimuli. The term is sometimes used in a broader sense to refer to the mental process by which all kinds of data—intellectual and emotional, as well as sensory—are organised meaningfully. See also Apperception.

Perplexity: A state of puzzled bewilderment in which verbal responses are desultory and disjointed, reminiscent of confusion. Its clinical associations include acute *schizophrenia, severe anxiety, manic-depressive illness* and the *organic* psychoses with *confusion.*

Phenomenology: The study of events or happenings in their own right, rather than from the point of view of inferred causes.

Pressure of speech: An increase in the amount of spontaneous speech; rapid, loud, accelerated speech. It is also called pressured speech. Occurs in mania, schizophrenia, and organic disorders. See also Communication disorder, Logorrhea.

Rapport: Conscious feeling of *harmonious* accord, sympathy, and mutual responsiveness between two or more persons. Rapport contributes to an effective therapeutic process in both group and individual settings. See also Countertransference, Transference.

Resistance: A conscious or unconscious opposition to the uncovering of unconscious material. Resistance is linked to underlying psychological defense mechanisms against impulses from the id that are threatening to the ego.

Stupor: State of decreased reactivity to stimuli and less than full awareness of one's surroundings.

Tangentiality: A disturbance in which the person replies to a question in an oblique, digressive, or even irrelevant manner and the central idea is not communicated. The term has been used roughly synonymously with loosening of associations and speech derailment, but in DSM-III it refers only to answers to questions and not to spontaneous speech. Failure to communicate the central idea distinguishes tangentially from circumstantiality, in which the goal idea is reached in a delayed or indirect manner.

Therapeutic alliance: Conscious contractual relationship between therapist and patient in which each implicitly agrees that they need to work together to help the patient with his problems. It involves a therapeutic splitting of the patient's ego into observing and experiencing parts. A good therapeutic alliance is especially necessary for the contribution of treatment during phases of strong negative transference. See also Working alliance.

Word salad: An incoherent, essentially incomprehensible mixture of words and phrases commonly seen in far-advanced cases of schizophrenia. See also Incoherence.

4

Psychoses: Schizophrenia, Brief Psychotic Disorder and Delusional Disorder

Psychotic disorders are serious mental illnesses in which the ability to recognise and respond rationally to reality, is lost. In clinical practice this manifests as odd or unusual thinking, perceiving and behaving. Psychotic disorders are commonly divided into three main groups—those of organic origin (where etiology is recognised to be due to an identifiable physical cause), disorders primarily of thought and behaviour and those with primary disturbance of mood.

This chapter is a brief description of psychotic disorders primarily of disturbance of thought, perception and behaviour. Of the three disorders in the title, schizophrenia is the most common. Brief psychotic disorder is rarer forms of the disease and will not be described further here. Delusional disorder is rarer than schizophrenia but commoner than Brief Psychotic Disorder. Its chief features are various types of delusions which occupy the sufferer's mind but do not disable him in daily life to the same extent as schizophrenia does. Hence, the degree of social and occupational disability is much less. Delusional disorder is one of "Schizophrenia Spectrum" disorders which means that it segregates in the same families as schizophrenia. These disorders will not be discussed further here.

Schizophrenia

Every day we read the term *"schizophrenic"* used in all contexts. This is a grave injustice to the enormity of the problem, and to the profound suffering associated with this disease. Relatives liken schizophrenia to living death, aptly also called *"cancer of the mind."* In India, three to four people in 10,000 fall ill with schizophrenia every year. One can therefore, estimate that there are approximately at least two million people suffering from schizophrenia in India at any point of time. In addition, over four lakh cases are added to this number every year. Nearly 40–60% of such sufferers end up significantly disabled. With increasing life span, these numbers are also increasing. Our ancients knew schizophrenia well. The Ayurveda called it *unmada* and described treatments for it—*Rauwolfia serpentina* or Sarpagandha, a drug still being used in modern medicine. Schizophrenia is an illness present in Indian society as well as all other societies down the ages.

Epidemiology

Schizophrenia occurs all over the world, and with similar symptoms and course. It is not a disease of any particular caste, creed or community, of the rich or the poor. It is found in all societies and geographical areas, and incidence and life time prevalence are roughly equal worldwide. However there is said to be a greater incidence in men than women, and in urban areas. This could also be due to social drift, because affected persons often leave home and migrate to urban areas.

Its prevalence rate is just below 1%, while disorders related to it (*"Schizophrenia spectrum"* disorders) are approximately 5%, while delusional disorders account for a fractional percentage of the general population.

It usually begins in adolescence just when the person is learning to be independent and to deal with the world. The majority of sufferers—men or women of any age group—continue to get episodes of illness; a minority never recovers from the first episode. Only the fortunate few get one episode and then recover partially or fully.

The course of Schizophrenia has been documented to be more severe in developed countries and the outcome better in developing and more rural societies, perhaps because of increased risk of substance abuse in West, and increased risk for suicidal and assaulting behaviour (10% of patients commit suicide).

Some Case Histories

Case 1

Shakuntala, in her twenties, suddenly became a very religious person, calling Shri Krishna her child and refusing to eat any food except what she herself offered as prasad, taken cold and only once a day. She believed that Krishna slept with her, so there was no question of any physical relationship with her husband. Shakuntala gradually stopped eating altogether and lay in bed. When the glass top of a table placed near her cracked in winter, she attributed it to Krishna writing his name on the table, and then disappearing with the glass. It was only when Shakuntala stopped bathing or changing her clothes and began to mutter to herself that the family knew that something was wrong, and brought her for treatment.

Case 2

Samir, a pavement seller, believed a Khan—a movie star—was out to kill him; in fact, the whole country was against him because he knew secrets against powerful people.

Case 3

Roshni believed her sons are not really her own but were planted on her during her delivery (possibly by the doctor, or possibly by her husband). She cannot explain the origin of her belief nor why anyone should substitute her children. This does not prevent her from looking after her children fairly competently, with help from her husband and in-laws!

Case 4

Neetu was a carefree, sociable, loving, fifteen year old, doing academically well in her private school. When she was in the tenth standard, her personality gradually began to change. She became restless and inattentive in class, moody, silent and withdrawn at home. Even more worrying to her parents, she started keeping herself as far away as possible from her father. One terrible day, the "truth" came out. She told her mother that when she was ten, her father had exchanged his daal bowl for hers, which meant he had put his sperms into her body. She had realised this truth only now.

This outlandish statement alarmed her mother and a psychiatrist was consulted. He prescribed antipsychotic drugs, which unfortunately did not suit Neetu. She refused to take medicines, and her behaviour became more and more queer. She now often laughed without reason, and sometimes would cry without reason too. She had to be withdrawn from school both because of her odd behaviour and also because of her repeated failures.

She became a recluse, hardly speaking to anyone at home, often not even bathing for days together. Finally, when she was 22, her parents heard of a new antipsychotic drug and consulted another psychiatrist. In addition to medication, rehabilitation and psychotherapy were also begun. The scars of schizophrenia will always remain on Neetu's psyche and she will never really be socially comfortable. But Neetu is now friendlier with her father, accepting that her delusions about him were not real. She cooks for her family once in a while, and is now thinking of joining Open School to complete her 12th.

Case 5

In a family, the disease may affect many members. Here is an example.

The first to fall ill was Laxmi Chand, at the age of 24. He began to feel that whenever he spoke to someone, his own brain would be uncovered and the other person could immediately understand what he was going to talk about. He felt he could no longer control his emotions. His speech became so disorganised even abusive, that the listener could not understand it. This went on for years. Now, he often complains of numerous bodily symptoms. He complains that whenever he speaks, he involuntarily switches on to a totally different topic and his brain is totally empty. No illness (except schizophrenia) has ever been diagnosed. He consults psychiatrists intermittently, is doing poorly at his job and finds it difficult to meet daily requirements of life.

His brother Suresh Chand fell ill at age 38. He isolated himself from his relatives because he believed that they had set evil spirits onto him. He became very religious, and spent all his time praying to try and drive away the evil spirits troubling him. These spirits control his body, mind and his thoughts even today, decades after the first onset. They pollute his thoughts, and put obstructions in his way. As a result he has become a total recluse.

Their sister Parvati was luckier. She fell ill relatively late, after her husband's death when she was 51. Five years before her illness began, she became reclusive and isolated. Then later on, she began to feel that many people were speaking inside her, commenting on her, and heard shrill voices, which closed her ears to all outside sound. They threatened her, and warned her that if she allowed anyone in her family to go out of the house the voices would kill them. As a result, she would not allow her children to go out at all.

Finally her daughter fell ill as early as 18 years of age. She felt fearful, someone was following her and was convinced all fruit sellers, sweepers and maids were after her. She felt some ghosts were using her body and spoiling her. These ghosts also took away her thoughts, and "gave" her strange tastes and smells. A famous movie star—xxx Khan though, favoured her and directed her to do different things via the TV.

Shakuntala, Samir, Neetu and Roshni are all suffering from schizophrenia. Apart from the dramatic symptoms outlined above, they had many other symptoms and disabilities.

Brief Psychotic Disorder • ICD-10: psychotic conditions with onset within 2 weeks and full remission within 1 to 3 months • DSM-IV TR: sudden onset of psychotic symptoms lasts 1 day or more but less than 1 month. Remission is full • *With/without marked stressors* • Emotional turmoil and confusion: good prognosis • Negative symptoms: poor prognostic features • Confusion or perplexity at the onset, • Good premorbid social and occupational performance • Absence of blunted or flat affect • *If the combined duration of symptomatology exceeds 6 months, then schizophrenia should be considered.*

Etiology

Schizophrenia has a very strong hereditary component. Familial risk has been documented through twin studies, which use two types of twin pairs for exploring the role of inheritance that is, monozygotic and dizygotic. These studies have shown that as relationship to the identified patient - "proband" - becomes closer, risk of schizophrenia increases. This risk increases from 1% in the general population to nearly 50% where both parents are schizophrenic or in pairs of monozygotic twins.

But even monozygotic twins are not always concordant (i.e., do not share the disorder). Also, despite strong evidence for genetic susceptibility, one or more specific gene/s has not been unambiguously identified so far. A given susceptibility gene may or may not result in the disorder. This ambiguity could be due to environmental influences or interactions with other genes. As an added complication, various combinations of genes may all lead to the same disorder.

This may be due, in part, to the critical role that the environment plays in modulating genetic susceptibility in mental disorders. Environmental factors such as gestational and birth complications, influenza epidemics or maternal starvation, rhesus (Rh) factor incompatibility may all end into the schizophrenia phenotype.

Alternately genetic susceptibility or "mistakes" in neurodevelopment during intrauterine life may also increase susceptibility to this disorder. Thus schizophrenia may be due to failure of the normal development of the brain or a disease process that alters a normally developed brain leading to multiple possible outcome types, subtle neurological manifestations, cognitive dysfunction and disturbances in affect.

The neurotransmitter hypothesis of schizophrenia postulates either increased or disturbed production or transmission of various neurotransmitters in the brain. The most researched among these are Dopamine, catecholamines and Gamma Amino Butyric Acid (GABA). Whether these disturbances are the cause or effect of the disease are still matters of debate.

Symptoms

Onset of illness may be sudden even catastrophic, or insidious and gradual. Sudden onset sufferers generally have a better prognosis.

In some people the symptoms may come and go only once, lasting each time for less than six months. Only last for a brief period, disappear, and appear again after some time. The illness may then stop recurring. This is called *schizophreniform disorder.*

In most people the disease runs a long and continuous course. The severity of symptoms and the functioning of a person may, however, wax and wane. The disease may erupt and become severe, but may again become placid. Some people, as they grow older, are fortunate to experience a gradual decline in symptoms. About 25 per cent people with schizophrenia become symptom-free in their later lives.

The illness is marked by a variety of symptoms. The most prominent features are: disordered thinking which becomes incoherent, disjointed and rambling; incongruity of emotion; impulsive actions and utterances; and hallucinations, where a person begins to hear voices, often of unfriendly kind, or sees objects that do not exist. Bizarre delusions is another common feature. Sufferers may also perform strange movements. Most people with schizophrenia do not recognise that their mental functioning is disturbed or that they need help. This lack of insight worsens their suffering because they do not understand the fact that medication is a necessity for them.

Symptoms of Schizophrenia

"Core" symptoms

Disorderly thinking

Incongruity of emotion

Impulsive actions and utterances

Bizarre delusions (with little or no appreciation of why the ideas are not acceptable to the people around).

Paranoid thoughts (the beliefs that one is surrounded by hostile forces which keep a close watch and secretly intervene to harm).

Hallucinations (hearing threatening or unfriendly voices, where none exist).

Passivity feelings (the person is convinced that his actions are controlled by an alien power).

Thought insertion and thought broadcasting (the person feels that his thoughts are not his own, that other people put thoughts into his mind or withdraw them and make them public in some way, that other people can read his thoughts, or that his thoughts are being said aloud or broadcast on radio or TV).

Incoherent bizarre behaviour

Incoherent, disjointed, or rambling speech

Abnormal posturing or movement

Difficulty in coping with home and work <http://www.ncbi.nlm.nih. goversentrezquery.fcgi? cmd=Retrieve&db=PubMed&list_uids=10655007&dopt= Abstract>-related responsibilities

Associated symptoms

Unexplained physical tiredness

Poor concentration

Sleeping problems

Appetite problems

Diminished sexual interest

Overly dependent behaviour

(In short, symptoms of almost all other mental illnesses in one form or other, at some time or other)

Difficulties in money management

Severe distrust or suspiciousness

Compromised learning ability

Poor memory

Physical violence

Risk of harming self

Poor grooming and hygiene

To develop a clearer understanding of the illness, it is important to look at its characteristic symptoms:

Delusions

Delusions are false ideas or beliefs that may sometimes be believed for short periods of time. However, their persistence and bizarreness eventually can only be explained on the basis of illness.

Delusions may be of grandeur, power or persecution, or even extremely bizarre. A patient complained of mobile jammers closing up his insides so that he could not pass stool! Another believed that strangers from space had passed some kind of rays into his brain because of which his brain stopped working.

She/He may also believe that others are controlling his thoughts or that his own thoughts are being broadcast to the world so that other people can hear them.

Delusional disorder: Presence of non-bizarre delusions lasting for more than 1 month without prominent hallucinations. Persecutory, Grandiose, Somatic type–Delusional parasitosis, Dysmorphic delusions, Body odour (Bromosis), Jealousy, Erotomanic/Amorous.

Hallucinations

People with schizophrenia may also experience hallucinations (false sensory perceptions), and may see, hear, smell, feel, or taste things that are not really there. Auditory hallucinations, such as hearing voices when no one else is around, are especially common: including two or more voices conversing with each other, voices that continually comment on the person's thoughts or behaviour, or voices that command the person to do something. Hallucinations must occur in clear consciousness.

Disorganised thinking and speech

Since speech is the mirror of thought, and thought processes are fragmented or disorganized, their speech may become incoherent or nonsensical. They may jump from topic to topic or string together loosely associated phrases. They may also combine words and phrases in meaningless way or make up new words. In addition, they may show "poverty" of speech (either less words or less ideas conveyed through words), or slow speech. They may suddenly stop talking in the middle of a conversation. In severe cases muteness—no speech at all—may intervene.

Bizarre behaviour

A person with schizophrenia may appear disheveled, oddly dressed (for example, wear multiple shirts, coats, scarves and gloves or use inappropriate make up), may talk to himself, may shout or swear without provocation, may walk backward, laugh suddenly without explanation, make strange faces, or may display clearly inappropriate sexual behaviour such as masturbate in public (extremely rare in our population). In rare cases, he may maintain a rigid, bizarre pose for hours, or may engage in constant, random or repetitive movements.

Social withdrawal

Several "negative" symptoms may occur—the most characteristic being social withdrawal—avoiding others, decreased emotional expressiveness, and talking in a low, monotonous voice, avoid eye contact with others, and display a blank facial expression. He may also have difficulty in experiencing pleasure and may not participate in any work or social activities. This lack of volition stops him from initiating and pursuing goal-directed activities.

Other symptoms

People with schizophrenia may face difficulties with memory, attention span, abstract thinking, and planning ahead. They commonly suffer from anxiety, depression, and suicidal thoughts. They may experience physical tiredness for no valid reason, may oversleep or find difficulty in sleeping, suffer a loss of sexual interest, become overly dependant, and face problems in money management.

Diagnosis

Schizophrenia is diagnosed by its psychopathology—abnormalities in thinking and perception as inferred by the person's speech and behaviour. Since there are no blood or laboratory tests:

a detailed history of mental and physical disorders in the individual and his family, and careful mental status examination, are essential for diagnosis.

Diagnosis of Schizophrenia: *Delusions, Hallucinations, Disorganised speech, Grossly disorganised/catatonic behaviour, Negative symptoms (affective flattening, alogia, avolition).*

Laboratory Test and Differential Diagnosis

Individual symptoms of schizophrenia can occur in a diverse number of illnesses–for example, those caused by head trauma or a brain infection. If a misdiagnosis occurs, it may result in incorrect or harmful treatment. Therefore, developing an independent diagnostic test, preferably laboratory based, is a priority research area for this disease.

Differential diagnosis • Drugs: of abuse, antihypertensives, NSAIDs, steroids, anti-depressants, anti-parkinsonian, and several others. • Brain injuries or disease: CV episodes, tumour, aqueduct stenosis, trauma. • Metabolic-systemic diseases: infectious, deficiency, syphilis, endocrine, autoimmune, AIDS. • Genetic-chromosomal phenocopies: Huntington's, chorea, Wilson's disease, XXX, XXY, XYY, familial basal ganglia calcification and others. • Mood disorders.

Investigations • Careful history and full physical-neurological examination • Complete Blood Count, Electrolytes, • Fasting glucose, • Lipid profile, • Liver, Renal and Thyroid function tests. • ECG and X-ray chest, head • EEG and MRI as needed • HIV and syphilis tests when needed.

Onset and Course of Illness

Western research describes men as having mostly unimodal onset of positive symptoms with peak incidence from 18 to 25 years of age. The onset of schizophrenia among women is said to be bimodal, with the first peak between 20 and 25 years of age and second peak after 40 years of age. The more acute the onset and later the age of onset, the better is the prognosis. Where the onset is insidious, there is poor intermediate and poor long-term course. The longer the duration of untreated psychosis, the worse is the outcome.

Course of illness is said to be more benign in females perhaps because Estrogen modifies course or because females show better response to antipsychotic drugs. The deficit form of schizophrenia with mainly negative symptoms is predominantly a male disease. Egocentric (Western) cultures traditionally are more stressful for men and here they show worse prognosis.

The disease progresses if left untreated. There is increased risk of self harm or suicide, or violence towards others. There is also higher risk of drug abuse, especially in the West and especially of nicotine. Up to 10% of patients commit suicide and up to one third attempt it at some point of the disease. Some may be totally withdrawn and uncaring of themselves and others. Many leave home and are lost and others remain disabled. Poor self care, inactivity and slowness become sources of distress to both sufferers and their families. Relatives are confused wondering if the person is ill or just pretending, as behaviour may swing from normal to abnormal.

This is not a self-limiting disease. Non-progression does not mean cure, as ups and downs may occur in the course of the disease. This illness is said to have a better outcome in developing rather than in industrialised countries. But even in developing countries up to half the sufferers may remain unemployed and dependent. The economic and human costs of this illness are high the world over.

Management and Treatment

In the acute stage—lasting usually for 4–8 weeks—the chief aim is to control acute psychotic symptoms. Uncooperative patients may need home visits and treatment in the home setting. They may also need to be brought to hospital either after preliminary treatment at home or (if dangerous to self or to others) under police escort. During the subsequent stabilisation phase, the aim is to consolidate therapeutic gains as risk of relapse increases if treatment is stopped, or there are various life stresses. Lastly during the stable or maintenance phase the patient is in remission, and the aim is to prevent relapse and improve level of functioning.

The mainstay of treatment in schizophrenia are drugs mainly antipsychotics. Newer antipsychotics are now the first drugs of choice. If there is no response, older antipsychotics can be tried. Among resistant cases Clozapine may be considered under strict monitoring of blood count parameters, as this drug can cause fatal leucocytopenia. Other modes of administration such as long acting injectable antipsychotics or mouth dissolving oral medication can also be tried. Individual selection is based on history of response to a drug, preference for a particular drug or mode of administration, and side effect profile of the drug. This is because most antipsychotics are broadly similar in efficacy and differ mainly in their side effect profiles.

Antipsychotic drugs antagonise postsynaptic dopamine receptors in the brain and are useful to manage acute positive psychotic symptoms, to diminish anxiety, induce remission, maintain the achieved clinical effect and to prevent relapse.

Some antipsychotic drugs available in India • Typicals: Chlorpromazine. Trifluoperazine, Haloperidol, Flupenthixol. • Atypicals/Newer: Thioridazine, Clozapine, Risperidone, Olanzapine, Quetiapine, Ziprasidone, Aripiprazole • Long Acting Injectables: Fluphenazine, Haloperidol, Flupenthixol, Clopenthixol, Risperidone.

Similar to other classes of drugs antipsychotics show several different types of side effects. Most common especially with older antipsychotics, motor side effects (table below) are seen while newer ones cause different metabolic side effects.

Motor side effects of antipsychotic drugs

Parkinsonian:	Tremor, rigidity, bradykinesia, festinant gait, rigid posture
Dystonias:	Tongue protrusion, ophisthotonus, spasmodic torticollis, oculogyric crisis
Akathisia:	Movement, restlessness
Tardive dyskinesia:	Abnormal involuntary movements of face, limbs, respiratory muscles.

Other common side effects (newer antipsychotics)

Sedation, Weight gain, hyperlipidemias and new onset type II diabetes mellitus; Increased risk for cardiovascular disease.

Outcome

Therapeutic advances over the last four decades have enabled most persons with schizophrenia to live in the community. Nevertheless the majority will continue to experience various symptoms and to have social and cognitive disabilities.

In several World Health Organisation studies on outcome, the most important predictors of good outcome were origin from a developing centre and acuteness of onset. Other predictors of good outcome were: being married, of female gender, better adjustment in adolescence, more frequent contacts with friends and no use of street drugs.

Studies state that 55% of sufferers show moderately good outcomes and 45% have more severe outcomes. Effective and earlier initiation of pharmacological treatment makes considerable difference in early course and a modest impact on long-term course. During the first 5 to 10 years patients may have multiple exacerbations followed by a Plateau phase where there is underlying deterioration. Later on, patients may regain some degree of social and occupational competence and become less disruptive and easier to manage.

The prophylactic use of antipsychotics reduces relapse rates. It is essential to continue pharmacological treatment because stopping medication often precipitates relapse. With each episode, patients seem to lose something that they do not recapture. Each successive episode requires longer returning to a remitted state. Hence multiple relapses are detrimental to overall outcome.

Role of Psychotherapy, Family Therapy and Vocational Rehabilitation

Rather than formal psychotherapy, supportive psychotherapy has a definite role to play to improve symptoms and enhance quality of life. It also helps to reduce disinterest and passivity. Psychotherapy should focus on education about illness, strategies for coping and social skills training. The relationship between social competence and psychiatric disorder has been known for quite some. In general studies clearly demonstrated the effectiveness of social skills training procedures in improving specific behavioural social skills as well as more global measures.

Reducing rather than increasing, the pressures and expectations that patients impose on themselves and those which others impose on them improves performance. The technique of paradoxical intention may be useful. This involves setting a target well beneath a patient's capability in order to enable him to attain it. Use of structured but realistic activity schedules and programmes, considerable amounts of patience and perhaps most importantly the ability to allow the patient to proceed at his own pace would be the best strategy for success.

Families are important allies of professionals in the management of schizophrenia. Educational and supportive family interventions have an important effect on relapse prevention. Working with the entire family from time to time helps both patients and careers and helps to prevent negative familial affect (criticism, hostility, over involvement). It also helps to reduce heavy burden of caring for chronically ill.

Family advocacy and mental health consumer movement has been especially active among sufferers and carers of schizophrenia. Such groups now growing in India too support members, provide education, dispel myths about the illness and improve treatment standards. They have played an important role in improving health care conditions for the mentally ill as well as for reducing the stigma associated with mental illness.

Vocational rehabilitation is the most important aspect of psychological intervention in schizophrenia. Supported employment programmes, individual placement programmes that immerse and support patients at the job site are the need of the hour in India. This is because schizophrenia patients have a low level of job retention both due to the disease process itself as well as due to stigma.

5

Mania and Bipolar Affective Disorder

OUTLINE

- Introduction
  - History
  - Definitions
- Epidemiology
- Etiology
- Classification and diagnostic criteria
  - CD-10 criteria
  - DSM-IV TR criteria
    - Bipolar I disorder
    - Bipolar II disorder
- Course
  - Morbidity/mortality
- Consequences of untreated
- Differential diagnosis
- Assessment
- Investigations
- Management
  - Acute episode
  - Hospitalisation
  - Acute hypomanic episode
  - Acute manic/mixed episode

Depressive episode
Prophylaxis
Psychotherapy
Psychoeducation
Out-patient follow up
Relapse prevention
Specific issues
Suicidality
Concurrent medical problems
Special patient groups
Prognosis

Introduction

History

Bipolar affective disorder has been known since ancient times. Greek thinking reflected 'humoral' theories, with melancholia/depression caused by excess of 'black bile' and mania by excess of 'yellow bile'. In 19th century Jules Baillarger described *'la folie a double forme'* and Jean Pierre described *'la folie circulaire'*, and Emil Kraepelin comprehensively described 'manic depressive insanity'.

It is one of the most common, severe and persistent psychiatric illnesses. Classically, periods of prolonged or profound depression alternate with periods of excessively elevated and/or irritable mood (known as mania). Symptoms vary between patients and in between episodes in same patient. Attention has to be paid both to patients' and third party information from family and friends.

Definitions

Euthymia: A normal positive range of mood states implying the absence of depression or elevated mood.

Dysphoria: An emotional state characterised by anxiety, depression, or unease. A state of feeling unwell or unhappy.

Dysphoric mania: is "prominent depressive symptoms superimposed on manic psychosis."

Expansive mood/affect: Expression of one's feelings without restraint, frequently with an overestimation of one's importance. Often associated with grandiosity.

Euphoria: Intense elation with feelings of grandeur.

Depressive episode: see diagnostic criteria.

Mania/manic episode: see diagnostic criteria.

Hypomania/hypomanic episode: see diagnostic criteria.

Cyclothymic disorder: A chronic disorder, with duration of 2 or more years. Numerous hypomanic and minor depressive episodes, with few periods of euthymia (i.e., never symptom-free for more than 2 months.)

Mixed episode: Occurrence of both manic/hypomanic and depressive symptoms in the same episode, every day for at least 1 week (DSM-IV TR) or 2 weeks (ICD-10).

Seasonal affective disorder: A variant of bipolar disorder, with consistent seasonal pattern to occurrence of episodes (usually winter depression and summer/spring mania, but may be vice versa).

Rapid cycling: four or more episodes (of any type) in a year.

Epidemiology

- Lifetime prevalence: 0.3–1.5% (0.8% bipolar I, 0.5% bipolar II);
- Male = Female, (bipolar II and rapid cycling more common in females);
- Mean age of onset 17–25 years,
- Age range 15–50+years (peaks at 15–19 years and 20–21 years; mean of 21 years).

Etiology

Factors identified as important include:

Genetic

- First degree relatives: 10–15% risks.
- MZ twins: 33–90% concordance,
- DZ twins: 23%.
- Children of one parent have a 50% chance of psychiatric illness.

Genetic liability appears shared for schizophrenia, schizoaffective and bipolar illness. Mode of inheritance remains unknown.

Biochemical

- Dysregulation of neurotransmitters at brain synapses: Noradrenaline (NA), serotonin (5HT) and dopamine (DA) all have been implicated.
- Abnormalities of biogenic amine metabolites.
- Neuroendocrinal dysfunction (HPA axis): Given the effects of environmental stressors and exogenous steroids, role has been suggested for glucocorticoids and other stress related hormonal responses.

Psychosocial

- Stressful life events (major loses, disappointments, deaths, divorces).
- Pre-morbid personality types that use internalising rather than externalising defense mechanisms.

- Negative cognitive distortions about self, environment, and life experiences.
- Learned helplessness.

Psychodynamic

- Loss of libidinal object
- Introjection of an ambivalently loved object.

Etiological theories

Kindling: The older hypothesis that suggests a role of neuronal injury through electrophysiological kindling and behavioural sensitisation similar to epilepsy.

Abnormal apoptosis: Abnormal programmed cell death in critical neuronal network controlling emotions.

Classification and Diagnostic Criteria

ICD-10 criteria: Requires at least two episodes, one of which must be hypomanic, manic or mixed.

DSM-IV TR criteria: Allows a single manic episode and cyclothymic disorder to be classified as part of bipolar disorder. Defines two subtypes:

Bipolar I disorder: One or more manic or mixed episodes **with or without** history of one or more depressive episodes.

Bipolar II disorder: One or more depressive episodes plus one or more hypomanic episodes **without** manic or mixed episodes.

Manic episode (A) A distinct period of abnormally and persistently elevated, expansive, or irritable mood, lasting at least 1 week (or any duration if hospitalisation is necessary) (B) During the period of mood disturbance, three (or more) of the following symptoms have persisted (four if the mood is only irritable) and have been present to a significant degree: 1. Inflated self-esteem or grandiosity; 2. Decreased need for sleep (e.g., feels rested after only 3 hours of sleep); 3. More talkative than usual or pressure to keep talking; 4. Flight of ideas or subjective experience that thoughts are racing; 5. Distractibility (i.e., attention too easily drawn to unimportant or irrelevant external stimuli); 6. Increase in goal-directed activity (either socially, at work or school, or sexually) or psychomotor agitation; 7. Excessive involvement in pleasurable activities that have a high potential for painful consequences (e.g., engaging in unrestrained buying sprees, sexual indiscretions, or foolish business investments) Diagnostic and statistical manual of mental disorders. 4th ed. Washington, D.C.: American Psychiatric Association, 1994 : 327, 332.

Summary of Manic and Depressive Symptom Criteria in DSM-IV-TR Mood Disorders

Disorder Manic Symptom Criteria Depressive Symptom Criteria

Major depressive disorder; No history of mania or hypomania; History of major depressive episodes (single or recurrent); Dysthymic disorder; No history of mania or hypomania; Depressed mood, more days than not, for at least 2 years (but not meeting criteria for a major depressive episode) Bipolar I disorder; History of manic or mixed episodes; Major depressive episodes typical but not required for diagnosis; Bipolar II disorder; One or more episodes of hypomania; no manic or mixed episodes; History of major depressive episodes; Cyclothymic disorder; For at least 2 years, the presence of numerous periods with hypomanic symptoms; Numerous periods with depressive symptoms that do not meet criteria for a major depressive episode. Bipolar disorder not otherwise specified. Manic symptoms present, but criteria not met for bipolar I, bipolar II, or cyclothymic disorder. Not required for diagnosis *Adapted from American Journal of Psychiatry 159:4, April, 2002.*

Course

It is extremely variable. First episode in females tends to be depressive and in males tends to be manic, but may be of any type. Untreated patients may have more than 10 episodes in a lifetime. Duration and period of time between episodes stabilises after 4th or 5th episode as many as 60% of people diagnosed with bipolar I disorder experience chronic interpersonal or occupational difficulties and subclinical symptoms between acute episodes. Environmental and lifestyle features can impact on severity and course. Stressful life events, changes in sleep-wake schedule, and current alcohol or substance abuse may affect the course and lengthen the time to recovery.

Morbidity/Mortality

High (in terms of lost work, productivity, quality of life and relationships) 25–50% attempt suicide and 10%–15% complete. Pharmacotherapy may substantially reduce the risk of suicide. Comorbidity is significant and increases suicidal risk—especially drug/alcohol misuse and anxiety disorder.

Consequences of Untreated

- High % have alcohol/substance dependence
- Suicide (up to 10% of bipolar patients commit suicide)
- Employment problems, financial problems
- Poor health, poor quality of life
- Marital disharmony, interpersonal problems

Differential Diagnosis

According to the episode

(A) Hypomania/Mania/Mixed Depression (*i*) ADHD (Attention Deficit Hyperactivity Disorder), (*ii*) Compulsive Disorders (Shoplifting, etc.), (*iii*) Impulse Control Disorders (*iv*) Schizoaffective disorder (*v*) Schizophrenia, (*vi*) Agitated depression (*vii*) Personality Disorders, (*viii*) Substance Intoxication (stimulants hallucinogens, opiates) **(B) Psychiatric disorders** (*i*) Dysthymia, (*ii*) Stress related disorders, (*iii*) Bipolar disorder, (*iv*) Anxiety disorders (obsessive compulsive disorder, panic, phobias), (*v*) Schizoaffective disorder, (*vi*) Schizophrenia (negative symptoms), (*vii*) Personality disorders (esp. Borderline personality disorder) (*viii*) Drug withdrawal: (Amphetamine, cocaine) **(C) Secondary Mania/Depression** Tertiary syphilis, influenza, AIDS, viral pneumonia, viral hepatitis, TB, infectious mononucleosis, **(D) Infection** (*i*) Tertiary syphilis, (*ii*) influenza, (*iii*) AIDS, (*iv*) viral pneumonia, (*v*) viral hepatitis, (*vi*) TB, (*vii*) Infectious mononucleosis, (*viii*) Multiple sclerosis, (*ix*) Head injury, (*x*) Brain tumour, (*xi*) Epilepsy, (*xii*) Encephalopathies e.g. HIV, (*xiii*) Neurosyphilis, **(E) Neurological disorders** (*i*) Dementia, (*ii*) Parkinson's, (*iii*) Huntington's, (*iv*) Multiple sclerosis, (*v*) Head injury, (*vi*) Brain tumour, (*vii*) Epilepsy, (*viii*) Encephalopathies e.g. HIV, (*ix*) Neurosyphilis, (*x*) Hypo/hyperthyroidism, (*xi*) Cushing's syndrome. **(F) Endocrine disorders** (*i*) Hypothyroidism, (*ii*) Apathetic hyperthyroidism, (*iii*) Diabetes, (*iv*) Hyperparathyroidism, (*v*) Late luteal phase dysphoria, (*vi*) Cushing's syndrome, (*vii*) Adrenal insufficiency, (*viii*) Hypopituitarism, **(G) Drugs** (*i*) Corticosteroids, (*ii*) Contraceptives, (*iii*) Antihypertensive (beta blockers, reserpine), (*iv*) Alpha methyldopa, (*v*) Anticholinesterases, (*vi*) Anticancer, (*vii*) Antiulcer (cimetidine, ranitidine), (*viii*) Indomethacin, (*ix*) Antipsychotics (*x*) Antidepressants, (*xi*) Psycho stimulants, (*xii*) Antibiotics, (*xiii*) Steroids etc.

Assessment

Full assessment is sought to take into account—

- Number of previous episodes
- Average length of episodes
- Level of psychosocial functioning in between episodes
- Previous response to treatment
- Family history of psychiatric problems
- Current and past use of alcohol and other drugs

Investigations

According to the episode—directed towards

- Identifying and/or excluding any organic/treatable cause, and
- As baseline before starting drugs.

Investigations—For organic/treatable cause

According to suspected cause—includes

- *Blood chemistry* e.g., for infectious etiology manifesting with significant physical symptoms resembling agitation or increased activity of mania.
- *Hormonal assays* e.g., for mood symptoms in a patient with signs of hypo/hyperthyroidism.
- *Electrolytes* e.g., for confusional states presenting similar to manic/catatonic excitement or depressive stupor.

- *Brain imaging*: e.g., CT or MRI scan for suspected dementia presenting as mood symptoms (e.g. disinhibition or pseudo-depression)
- *Serology*: e.g. testing for VDRL or HIV.

Investigations—For Medicines

Investigation **Lithium Carbamazepine Valproate** CBP (Complete blood picture)? (May cause leucocytosis)? (May cause bone marrow depression)? (May cause benign thrombocytopenia) RFT (Renal Function Tests)? (May cause nephrogenic diabetes insipidus) LFT (Liver Function Tests)? (May cause hepatotoxicity) ECG (Electrocardiogram)? (May cause "sick sinus syndrome") Electrolytes (Na, K, Cl, Ca)? (May cause hyponatremia)? UPT (Urine Pregnancy Test)? (For women of reproductive age group) TFT (Thyroid Function Tests)? (May cause hypothyroidism) **Laboratory Monitoring** Monitoring **Lithium CBZ Valproate First 2 months of therapy** Weekly/2 Weekly (Serum level) Monthly CBC (Complete blood counts) and liver function tests CBC and liver function tests **Long-term therapy** Every 3 to 6 months (Serum level) Every 6 months Thyroid function tests yearly (total T4, T4 uptake and TSH) CBC and liver function tests every 6 months Every 6 to 12 months Renal function test. CBC and liver function tests every 6 to 12 months.

Management

Acute Episodes

General Plan

- Developing of an effective therapeutic alliance with the patient and his/her family and friends
- Decide the Locus (location), FOCUS (priority symptoms) and MODUS (pharmacotherapy/psychotherapy)

Locus (site-outpatient/inpatient(IP)-closed or open ward).

Principle-Manage in the Least Restrictive Setting

Outpatient Home if no indication for IP treatment Day care/community. As transition from IP or if appropriate Inpatient Open ward. Low risk for Dangerous behaviour(s). Closed ward High risk for Dangerous behaviour(s).

Focus (of immediate attention)

- Control of severe agitation and violence
- Suicidal (Expressing ideas/plans/Attempted recently/Psychomotor excitation ++/ Hopelessness ++
- Gross nutritional disturbances
- Any serious general medical condition
- Any serious side effect of treatment

Modus

- Somatic methods of treatment—Psychotropics and/or ECT
- Psychoeducation

Hospitalisation

Indications for Hospitalisation:

- Clinical symptoms and situations where admission may be necessary.
- High risk of harm to self (suicidal), or to others (agitated/violent/homicidal)
- Behaviour (illness related) that endangers relationships, reputation or assets
- Poor self care/gross neglect
- Serious/Disabling side effects
- Poor social support
- Severe symptoms—Depressive/Manic/Mixed
- Psychotic or Catatonic
- Rapid cycling
- Failure of outpatient treatment
- Rationalisation of treatment/medicines

Acute Episodes

Specific Plan

Acute Hypomanic Episode

Without a history of rapid cycling, use a "traditional mood stabiliser" (Lithium, Valproate and Carbamazepine).

Acute Manic/Mixed Episode

FIRST LINE OF TREATMENT –

Mood Stabilisers

Lithium –

- Best evidence
- May require up to 2 weeks for effectiveness – usually requires addition of antipsychotics/ benzodiazepines.
- Predictors of good response: < 3 episodes, Classic bipolar I symptom pattern: euphoric mania and severe depressions, and absence of substance abuse and psychotic symptoms, past or family h/o good response to lithium.
- Cheap, inexpensive drug.
- Antidepressant action. Helps reduce suicide risk.
- Start at dose of 300 mg (plain preparation) t.i.d. for average male, test serum levels after 5 days and adjust to maintain levels of – 0.8–1.2 meq/dl (for treatment of acute episode, 0.6–1.0 meq/dl(for prophylaxis).

Drug Interactions with Lithium

Drug effect on lithium level management: ***Thiazide diuretics***—Increased lithium level; Avoid this combination or reduce dosage; monitor lithium level; ***Loop diuretics***—Avoid this combination or alter either dosage as needed; monitor lithium level; ***Potassium-sparing diuretics***—Decreased lithium level monitor lithium level and adjust dosage. ***Nonsteroidal anti-inflammatory drugs***—Increased lithium level or Use lower dosage of lithium; consider aspirin or sulindac; ***Angiotensin-converting enzyme inhibitors***—Increased lithium level; toxicity reported; Use lower dosage of lithium; monitor lithium level closely; ***Calcium channel blockers***—Increased or decreased lithium level; Monitor lithium level closely.

Valproate

- Well tolerated with very few drug interactions.
- Can be started at loading dose of 20 mg/kg body weight, therefore better for rapid control of severe symptoms.
- Predictors of good response-dysphoric mania, mixed episode, rapid cycling, presence of substance abuse or psychotic symptoms.
- Costliest among the first line drugs.
- Weight gain risk.

Drug Interactions with Valproic Acid

Drug interaction management: ***Phenobarbital***—Increased phenobarbital level: Reduce dosage. ***Magnesium and aluminium-containing antacids***—Increased valproic acid level or Monitor valproic acid level; reduce dosage ***Carbamazepine***—Decreased valproic acid level; possible increased carbamazepine level; Monitor valproic acid level; adjust dosage ***Aspirin and naproxen***—Increased valproic acid level; Avoid salicylates or other drugs bound to plasma albumin. ***Benzodiazepines***—Increased sedation; Use with caution.

Carbamazepine

- Has to be titrated up slowly—regime of dose starting and increment similar to antiepileptic regimes.
- Higher incidence of side effects.
- Cheap drug with good efficacy.
- Predictors of good response dysphoric mania, mixed episode, rapid cycling, and presence of substance abuse or psychotic symptoms, severe sleep problems.
- Low risk of weight gain.

Antipsychotics

- Useful for rapid control of severely agitated or psychotic patients.
- Use of atypical antipsychotics preferred (e.g. Risperidone 4–6 mg/Olanzapine 10–20 mg.)

- Olanzapine good for rapid control of severe symptoms, shown to be an effective mood stabiliser with few short-term risks.
- Clozapine effective for treatment resistant cases, but requires weekly blood tests.
- And risk of weight gain.
- Aripiprazole and Ziprasidone "weight neutral".

Seen to have both antimanic and antidepressant effects-Lithium <http://www.psycheducation.org/depression/meds/lithium.html>, Olanzapine <http://www.psycheducation.org/depression/meds/olanzapine.htm> and Lamotrigine <http://www.psycheducation.org/depression/meds/lamotrigine.html>.

Can prevent recurrences – Lithium <http://www.psycheducation.org/depression/meds/lithium.html>, Valproate, Carbamazepine, Olanzapine <http://www.psycheducation.org/depression/meds/olanzapine.html>, Lamotrigine <http://www.psycheducation.org/depression/meds/lamotrigine.html>, and Aripiprazole <http://www.psycheducation.org/depression/meds/2ndGens.html>

Bipolar Depressive Episode

Treatment by mood stabiliser alone, or/and antidepressants, ECT

- Risk of precipitating (hypo) manic/mixed episode or "Switch"
- Severely depressed/high suicidal risk/needing urgent treatment—ECT may be considered first line
- Drug free patient – start mood stabiliser (e.g. lithium or lamotrigine with antidepressant properties), or antidepressant (with lesser risk of inducing switch, e.g. Bupropion, Paroxetine).
- Already on prophylaxis – optimise drug (check compliance, check serum levels), treat/exclude comorbidity that might affect treatment (e.g. drug/alcohol abuse, hypothyroidism)
- Add additional mood stabiliser or antidepressant if need.
- In general, the experts rely heavily on lithium and lamotrigine; and use caution with antidepressants (concern for the risk of inducing hypomania and mania; and in recognition of lack of data supporting antidepressants in this role, versus lithium alone).

Prophylaxis

- Aims to prevent recurrent episodes
- Indicated in any patient with 2 or more episodes in past 5 years
- Lithium or Valproate may be the first line of treatment
- Newer antipsychotics may be used to augment, especially in patients with psychotic symptoms.

Psychotherapy

Most of the therapies emphasise similar ingredients:

- Identifying signs of relapse, making plans for early detection and response.
- Using education to increase agreement between doctor, patient and family about what it being treated and why;
- Emphasis on the need to stay on medications even when well;
- Stress management, problem-solving, and focus on improving relationships; and
- Regular daily "rhythms" for sleep, exercise, eating, activities.

Therapies that have been tested are: Prodrome Detection <http://www.psycheducation.org/depression/Psychotherapy.html>, Psychoeducation <http://www.psycheducation.org/depression/Psychotherapy.html>, Cognitive Therapy <http://www.psycheducation.org/depression/Psychotherapy.html>, Interpersonal/Social Rhythm <http://www.psycheducation.org/depression/Psychotherapy.html>, Family-Focused Therapy <http://www.psycheducation.org/depression/Psychotherapy.html>, Support groups.

Set limits on impulsive behaviour in patients with mania.

Hold family meetings to discuss issues.

Psychoeducation

Targets are:

- Regulating social and biorhythms (Sleep: Lack of sleep can provoke a hypomanic or manic episode)
- Avoiding or regulating alcohol or substance use (Substance use may exacerbate a mood disorder, particularly rapid cycling)
- Proactively dealing with interpersonal conflict, high expressed negative emotions.
- Specific strategies to monitor moods, reduce or contain suicidality, and improve medication adherence all promote a better prognosis
- Negative attitudes towards medication in the patient, key family member or friend, or a member of the health care team.

Outpatient Follow-up

Aims of OPD follow-up –

Establish and maintain a therapeutic alliance

Monitor

- Psychiatric status [symptoms of BAD (e.g. suicidality, mood), sleep patterns, life events, substance use, activity and any co-morbidities].
- Medical status (any co-morbidities)
- Compliance (by asking family members and by doing blood levels of mood stabilisers)

- Side effects of medicines (by monitoring signs, symptoms of Toxicity) and
- Blood levels of mood stabilisers

Psychoeducation about

Disease –

- Increase understanding of signs, symptoms risks and biologic nature of the illness
- Importance of compliance with therapy.
- Identify new episodes early (identify early signs, symptoms)
- Increase understanding of and adaptation to psychosocial effects
- Management of stress and management of work and leisure activities.
- Limited caffeine and alcohol intake and

Drugs –

- Dosage, duration, effects, side effects and toxicity.
- Improve compliance

Regularity – in rhythms of activity and wakefulness, sleep hygiene, eating and exercising regularly.

Management of long-range issues that may include marital problems, employment and financial problems, peer relationships and modification of personality traits.

Relapse Prevention and Psychoeducation

Helping patients to identify precipitants or early manifestations of illness, so that treatment can be initiated early, is a key part of psychiatric management.

Early or subtle signs of mania or depression should be recognised and treated (mania/mixed episodes with short-term use of antipsychotics and depression with increased doses of antidepressants).

Education about the "S" that may precipitate episodes should be given—

Sleep (sleep deprivation may precipitate mania and excess of sleep may lead to depression),

Season (depression may occur seasonally in winters, and mania in summers/spring), and

Stress (may precipitate any type of episode).

Specific Issues

Suicidality

Characteristics to Evaluate in an Assessment of Suicide Risk in Patients with Bipolar Disorder

- Presence of suicidal or homicidal ideation, intent, or plans
- Access to means for suicide and the lethality of those means
- Presence of command hallucinations, other psychotic symptoms, or severe anxiety
- Presence of alcohol or substance use

- History and seriousness of previous attempts
- Family history of or recent exposure to suicide

Concurrent medical problems

Assess and treat accordingly, as untreated might affect natural course of illness, and the management.

Special patient groups

Children and adolescents: Mania may be more chronic, difficult to treat and may require multidrug therapy.

Elderly: Onset of mania after age 60 is less likely to be associated with a family history of bipolar disorder and is more likely to be associated with identifiable general medical factors, including stroke or other central nervous system lesion.

Pregnancy and lactation: ECT may be considered as first line as it helps in faster recovery.

Prognosis

Poor prognostic factors: drugs/alcohol abuse, psychotic features, non-compliance, poor employment history, male sex, residual symptoms between episodes, long duration of episodes.

Good prognostic factors: later age of onset, few co-morbid physical problem, good treatment response and compliance, full recovery between episodes, manic episodes of short duration.

Case Vignette

Mrs. S.K. was a married lady of 27 years having two children with a *family history of bipolar disorder in her father*. In the past, at the age of 24 years, she had an episode of depression after the delivery of her second baby. She was taken to the hospital because she was very *excited and talkative* for 2 weeks. She was very *agitated, could not sleep, talked almost incessantly and refused her food*. Her *endless conversation* was mainly about films, actors and she interrupted it only to *sing and dance.*

The doctor observed that Mrs. S.K. was tidily, even smartly, dressed. She appeared excited and irritable, with aggressive shouting. She was very talkative, and her speech was sometimes difficult to follow because she spoke very quickly, jumping from one topic to another. She felt superior to others, who were jealous because of her voice and beauty. She was easily distractible, but oriented to time, place and person. She did not show any impairment of memory or other cognitive functions. Physical and neurological examinations and laboratory investigations, including thyroid parameters, were all normal.

Her diagnosis is Bipolar Affective Disorder, current episode mania.

6

Depression in General Practice

Introduction

Depression is one of the commonest psychiatric disorders. Most of the patients often present in primary care settings rather than to a psychiatrist. A major proportion remains undetected leading to chronicity and disability. It is a major public health problem and has also been identified as a leading cause of disability in the recent World Health Organisation—World Bank study of Global Burden of Disease.

Depression or depressed mood is a normal emotional experience in everyday life. We all become depressed at one time or other in everyday life depending on the circumstances. However, once this emotional state becomes unusually long, it could be abnormal. Depressed mood, when persistent and pervasive, may be accompanied by a number of signs/symptoms and then becomes an illness i.e., depressive disorder.

Historically a number of different terms have been used in relation to the depressive illness. These include manic depressive illness, manic depressive psychosis (MDP), psychotic and neurotic depression, endogenous and reactive depression, unipolar and bipolar depression and melancholia. Most of these now have more or less historical significance.

In the current classificatory systems, depressive disorders are categorised as depressive episode (when it is the first episode), recurrent depressive disorder, depressive episode occurring in bipolar affective disorder and dysthymia. The individual episodes are further classified as mild, moderate and severe, depending on severity.

Epidemiology

Lifetime prevalence of major depression is 10–25% in women and 5–12% in men with one year prevalence of 5%. Minor depressive disorders are equally common.

Most patients present in primary care with somatic symptoms. A large number of cases remain unrecognised.

Aetiology

Deficiency of neurotransmitters like NE, serotonin, and dopamine in limbic system is responsible for the illness. Antidepressants tend to correct the disturbance. In cases of chronic and recurrent depression, neuronal atrophy and cell deaths of specific neuronal populations have been found in hippocampus and prefrontal cortex. Chronic antidepressant treatment through regulation of cAMP Cascade could reverse or prevent such neuronal damage and atrophy of damaged or vulnerable neuronal populations.

Clinical Picture

Depression can occur at any age. Depressed mood is the central feature and all other symptoms are secondary to it. The patients more often present with somatic symptoms like aches and pains or pain affecting other body areas like head, back, neck or chest, lethargy, disturbances of appetite and sleep, which could easily be passed on as indicating a physical illness. Usually it is possible to elicit depressed mood or feelings of disinterest, if the clinician is observant and specific questions are asked. The chronic patients or patients suffering a relapse might present with the core depressive symptoms like feeling depressed and low, losing interest in day to day activities.

A pervasive depressed mood and disinterest or inabilities to experience pleasure in activities which one enjoyed earlier are the key symptoms. Both or either of these two symptoms should be present throughout the day on most of the days for at least 2 weeks along with other symptoms for making a diagnosis of depression.

Other symptoms include fatigue/loss of energy, changes in appetite mostly reduced, loss of weight (increased in few), sleep disturbances (difficulty in initiating sleep, frequent disruptions in sleep in midnight, early morning insomnia or increased sleep in a few), slowing of movements and thinking (psychomotor retardation) or psychomotor agitation, ideas of self reproach and guilt, impairment of concentration presenting as forgetfulness and ideas of hopelessness, helplessness, worthlessness, death wishes and suicidal ideation. There may be a history of over dosage or suicidal attempts in the past.

There is a definite change from normal functioning.

Depression is further categorised into mild, moderate or severe, depending on the severity of dysfunction and number of symptoms. Similarly, the moderate and severe depression are further categorised as with somatic or psychotic symptoms, depending on their presence.

Depression is an episodic illness. If recurrent episodes are occurring, it is called recurrent depressive disorder.

Differential Diagnosis

A number of prescription medications and common substances of abuse, and medical conditions are associated with depression. The primary care physician who is often dealing with physically ill patients needs to be observant about these causes of depression.

1. Drug Induced Depression

Some of the common prescription medications and substances of abuse known to cause depression include:

- **Antihypertensives:** Hydralazine, propranolol, reserpine, methyldopa
- **Sedatives and hypnotics:** Barbiturates, benzodizepines, opioids
- Alcohol
- Steroids
- Antineoplastics
- **Neurological agents:** Amantadine, bromocriptine, L dopa
- **Analgesics:** Ibuprofen, indomethacin
- Antibacterials

If the patient is on these medications and the depressive symptoms have started after these were prescribed, chances of these medications being responsible for depression are high. In such a situation, attempt should be made to shift to an alternative, if possible.

2. Medical Conditions Causing Depression

A number of medical conditions may present with depressive symptoms. Some of the common medical illnesses associated with depression are listed below:

Degenerative neurological conditions: Parkinsonism, Alzheimer's, Huntington's Disease

Cerebrovascular disease: Stroke

Metabolic: B_{12} deficiency

Endocrine: Hypo-and hyper-thyroidism, Hypo-and hyper-parathyroidism, hypo-and hyper-adrenocorticism

Auto-immune disorders: Systemic Lupus Erythematosus (SLE)

Viral and other infections: HIV, hepatitis

Malignancies: Carcinoma pancreas.

Other Depressive Disorders

Dysthymia

Dysthymia is a chronic depressive illness, in which the patient usually presents with depressed or irritable mood for many years. The patient more often feels tired and lethargic. Disturbance of sleep and appetite are often present. Symptoms are not severe enough to make a diagnosis of depression. Onset is usually in adolescence or early 20s. Symptoms should be present for at least 2 years before a diagnosis of dysthymia can be made.

Mixed Anxiety Depression

Mixed anxiety depression is more often seen in general practice rather than in psychiatric practice. Symptoms both of anxiety as well as depression are present, but each category not sufficient for the individual diagnosis. The patients frequently present with vague somatic symptoms along with a mixture of anxiety and depressive features.

Course and Prognosis

Depressive disorders often run an episodic course and tend to recur. After a single episode, 50–85% patients have a second relapse and 80–90% of those who have had a second episode would have a third episode. Up to 15% of depressives end their life by suicide. The drugs reduce the chances of relapse and improve quality of life and disease associated disability and burden.

Management

Depression can be easily identified and diagnosed, if the primary care clinician is observant and carries a high rate of suspicion especially when dealing with patients presenting with unexplained physical symptoms. Once diagnosed, it can be treated effectively. Both pharmacological as well as non-pharmacological approaches are used for treatment.

A. Pharmacological Management

A wide variety of effective drugs, but there is always a lag period of 2–3 weeks for response to start: Most have nearly equivalent efficacy, but differ in the side effect profile. Individual variations are observed in the response to medication. A list of antidepressants available currently in India is given below:

1. Heterocyclic Antidepressants: Imipramine, Amitriptyline, Nortriptyline, Doxepin, Dothiepin, Clomipramine, Mianserin, Amoxapine.
2. Serotonin Specific Reuptake Inhibitors (SSRIs): Fluoxetine, Sertraline, Paroxetine, Fluvoxamine, Citalopram, Escitalopram.
3. Serotonin and Norepinephrine Reuptake Inhibitors (SNRIs): Venlafaxine, Duloxetine.
4. Norepinephrine Reuptake Inhibitors (NRIs): Reboxetine.
5. Norepinephrine and Serotonin Specific Antidepressant (NSSA): Mirtazapine.
6. Reversible Inhibitors of Monoamine Oxidase (RIMAs): Moclobemide.
7. Serotonin-2 Antagonists. Trazadone, Nefazodone.

1. Heterocyclic Antidepressants

Heterocyclic antidepressants are the oldest group of antidepressants. The group includes the conventional tricyclics. The efficacy of this group of drugs is unquestionable. Tricyclics like imipramine and ammtriptyline were the most commonly used antidepressant till early 1990s. The starting dose of imipramine and amitriptyline is 75 mg in 2–3 divided doses, which is gradually built up to 150 mg in 7–10 days time. The maximum dose is 250–300 mg/day. The tricyclics have severe anticholinergic (dryness of mouth, blurring of vision, constipation, urinary retention etc.), antiadrenergic (postural hypotension), sedation and cardiac side effects, weight gain and sexual dysfunction.

Mianserin is a tetracyclic antidepressant used in dose of 30–60 mg/day usually as a single dose. Common side effects include sedation and giddiness. There have been reports of blood dyscrasias associated with the use of mianserin, which have reduced its use.

Amoxapine is another tetracyclic. One of the metabolites of amoxapine has antipsychotic properties and it has been promoted especially in cases of depression with psychotic features. The dose of amoxapine is 150–400 mg per day in divided doses.

2. Serotonin Specific Reuptake Inhibitors (SSRIs)

SSRIs are often the drugs of first choice for treatment of depression. Currently 6 SSRIs (fluoxetine, sertraline, fluvoxamine, paroxetine, citalopram and escitalopram) are available. Doses of different SSRIs are given in Table 1. Lowest dose is often the therapeutic dose and most of the patients respond to this dose. SSRIs have a long half-life and are therefore usually given as a single dose (except fluvoxamine, which is given in 2 divided doses), usually in the morning after breakfast because of their tendency to induce insomnia and gastric distress.

Table 1: Doses of SSRIs

SSRI	*Dose in mg per day*
Fluoxetine	20–80 mg
Sertraline	50–200 mg
Fluvoxamine	100–300 mg
Paroxetine	20–50 mg
Citalopram	20–50 mg
Escitalopram	10–20 mg

However, in severe depression SSRIs have a limited efficacy in severe depression.

- Common side effects include nausea, vomiting, dyspepsia, agitation, anxiety, headache and sexual dysfunction. Out of the various SSRIs, citalopram and escitalopram have minimal drug interactions and therefore are safer in patients on treatment for physical problems. Escitalopram is an isomer of citalopram.

3. Serotonin and Norepinephrine Reuptake Inhibitors (SNRIs)

Venlafaxine and duloxetine are two SNRIs currently available in market. Venlafaxine is started in dose of 37.5 mg per day and gradually increased to 150 mg per day after 1–2 weeks. Maximum dose is 300 mg per day. The extended release preparations are also available for once a day used. Higher doses are associated with elevation of BP making it not a preferred drug for patients with hypertension.

Whenever a higher dose is to be used, BP should be monitored. Other side effects include nausea, insomnia, dry mouth, drowsiness, dizziness, headache, sedation, sexual dysfunction. Venlafaxine has also got efficacy in severe depression.

Duloxetine is a new introduction in the filed of antidepressants. It does not have significant cardiovascular side effects. It is used in the dose of 60–120 mg/day. Most patients respond to a dose of 60 mg per day, given as a single dose. Common side effects include nausea, insomnia, dry mouth, drowsiness, dizziness, headache and sedation.

4. **Norepinephrine Reuptake Inhibitors (NRIs)**

   Reboxetine is a member of this group of drugs. It is used in doses of 4–6 mg twice a day. Usual starting dose is 4 mg BD, which can be increase if required. Common side effects include insomnia, sweating, dizziness, dry mouth, constipation, tachycardia and urinary hesitancy.

5. **Norepinephrine and Serotonin Specific Antidepressant (NSSA)**

   Mirtazapine belongs to NSSA group. It is used in doses of 15–45 mg per day. It is usually used as a single bedtime dose. Usual starting dose is 15 mg per day. Many patients would respond to the lowest dose. It is also available in form of dispersible tablets. Common side effects include increased appetite, wt. gain, drowsiness, oedema, dizziness and headache. Out of various antidepressants, mirtazapine has minimal sexual side effects.

6. **Monoamine Oxidase Inhibitors (MAO Inhibitors)**

   MAO inhibitors have a limited use in the current psychiatric practice because of the serious food and drug interactions. These are also not available in India. The reversible inhibitors of MAO (RIMAs) are relatively safe. Moclobemide, a member of this group is used in doses of 150–600 mg per day in two divided doses. The MAO inhibitors and RIMAs are especially indicated in cases of non-responsive and atypical depression (patients of depression presenting with irritability, increased sleep and increased appetite). Common side effects include sleep disturbance, nausea, agitation and confusion. Hypertension has also been reported in some cases and may be tyramine related.

7. **Serotonin-2 Antagonists**

   Trazadone and nefazodone are members of this group. Trazadone is used in doses of 150–300 mg per day. It is better to start in dose of 25 mg twice a day and to gradually build up the dose as it can cause severe giddiness, if started in higher doses. Most patients develop tolerance to the side effects. Common side effects include sedation, dizziness, headache, nausea, vomiting, postural hypotension, tachycardia and priapism.

Nefazodone is currently not available in India. It is used in dose of 300–600 mg per day. The recommended starting dose is 100 mg twice a day. Common side effects include headache, dry mouth, sedation, nausea, dizziness, constipation, insomnia, weakness and light headedness.

Choice of Antidepressant

With a large number of antidepressants available for use, it is often difficult and confusing which antidepressant to choose. Some general guidelines are given below:

- In mild depression, it is better to wait and watch. A simple counselling, reassurance and use of benzodiazepines (diazepam 5–10 mg per day, Lorazepam 2–3 mg per day or alprazolam 0.5–0.75 mg per day in divided doses) for 2–3 weeks may be a better strategy. It is important to mention here that the benzodiazepines should not be used for 2–3 weeks. A preferred approach would be to start tapering off after 10–15 days.
- SSRIs are the first line of treatment, unless the patient has shown good response to another antidepressant in the past.
- While using SSRIs, be watchful in early period for restlessness, agitation and suicidality (especially in young people below 30).
- Mirtazapine is another good choice.
- TCAs and venlafaxine are indicated in severe depression.

Non Responding or Resistant Depression

If a patient fails to respond even after taking an antidepressant for 5 weeks in adequate dose, the following strategies may need to be used:

- Check if treatment has been given in adequate dose for adequate period
- If yes, change antidepressant, another SSRI or a drug from different group
- Augmentation strategies can be used in resistant depression (Lithium, triiodothyronine, olanzapine). It is better to refer the patient to a specialist at this stage.
- Electroconvulsive therapy (ECT) is used in patients of depression with suicidal risk, poor oral intake and psychotic symptoms.

Duration of Treatment

A single episode of depression needs to be treated for 4–6 months after recovery. If treatment is stopped immediately, 50% patients are likely to relapse. In cases of recurrent depression, the treatment needs to continue for period varying from 2 years to indefinitely. While stopping the medication at the end of treatment, the drugs should be gradually tapered of over a period of 1–3 months depending on the length of treatment. Another important point in pharmacotherapy is that the therapeutic dose and the maintenance dose are the same. The dose should not be reduced in the maintenance phase.

B. Non Pharmacological Management

Non pharmacological approaches for treatment of depression include non-specific approaches like supportive psychotherapy, psycho-education, counselling, and reassurance and specific therapies like Cognitive Behaviour Therapy (CBT) and Inter Personal Therapy (IPT).

1. Supportive Psychotherapy

Supportive psychotherapy includes the components of reassurance, encouragement, emotional support and guidance. The patient and family need to be educated about the illness. Often the patient as well as the family has notions of depression not being an illness, which needs to be corrected. Need of long-term requirement of the medication in the background of long-term nature of illness should be emphasised. Many patients have an apprehension that the antidepressants may not be safe on long-term use or they might become dependent on them. Such fears need to be allayed. Antidepressants don't have any dependence potential.

2. Cognitive Behaviour Therapy (CBT)

CBT is a specific form of psychotherapy, which was especially developed for depression, though it is now also used in a variety of psychiatric disorders. It is based on the principle that the depression and the depressed mood are a result of faulty cognitions or maladaptive ways of thinking. The treatment is focussed at correcting the faulty cognitions.

It is a structured treatment and has proven efficacy. It is usually taken in form of weekly sessions spread over 12–16 weeks. The sessions can be held in one to one format or in small groups of patients. The treatment involves helping the patient identifying depressogenic negative thoughts, challenging negative automatic thoughts and learning rational responses to negative thoughts. The efficacy is likely to last for long time and it also has prophylactic value.

3. Interpersonal Therapy (IPT)

IPT is based on the principle that depression occurs in interpersonal context. IPT is also a specific treatment for depression. There is often a background of interpersonal conflicts, interpersonal deficits, and grief or role transition. The therapy is aimed at resolving the underlying problem situation. Treatment is structured and goes on for 12–16 weekly sessions.

Conclusion

Depression is a very common illness in primary care settings. If not treated early and effectively, it is associated with increased health care costs, disability and burden. It can be easily treated at the primary care level and the primary care physicians are fully equipped to treat such patients. Timely diagnosis and treatment can prevent distress and disability.

7

Psychoactive Substance use Disorders

Almost all cultures have used psychoactive substances to facilitate social interaction, to alter consciousness, to heal. Our society's expanded chemical manipulation simply represents large technical capacity, more wealth, leisure and individual choice. Among many a ill that have plagued the human society today, the drug abuse is of greatest social relevance.

To begin with, let us understand what is a *drug*. World Health Organisation (WHO) defines a *drug* as any substance that, when taken into the living organism, may modify one or more of its functions. This definition conceptualises '*drug*' in a very broad way, including not only the medicines but also the other pharmacologically active agents.

Drug addiction is a term that was used earlier for drug-dependence, has been dropped from the scientific per se for its derogatory connotation. Other terms, currently in use are 'harmful use', and 'psychoactive substance use disorder'. A psychoactive drug is one that is capable of altering mental functioning.

There are four important patterns of drug use disorders, which may overlap each other. These are as follows:

1. Acute Intoxication
2. Withdrawal State
3. Dependence Syndromes
4. Harmful Use

    1. *Acute Intoxication* is a transient condition following the administration of alcohol or other psychoactive substance, resulting in disturbances in level of consciousness, cognition, perception, affect or behaviour, or other psychophysiological functions and responses. This is usually associated with high blood levels of the drug. Sometimes even a low dose of the drug or alcohol is sufficient to cause intoxication. This may occur either due to low threshold (i.e., chronic renal failure) or due to idiosyncratic

sensitivity. As the time passes, the intensity of intoxication lessens and the symptoms eventually disappear in the absence of further use of the substance. The recovery is generally complete except in cases where tissue damage or other complications have occurred.

2. *Withdrawal State* comprises a cluster of symptoms that are specific to the drug used and develops on total or partial withdrawal of the drug usually after repeated and/or high-dose use. Withdrawal syndrome is generally short lasting ranging from few hours to few days. Withdrawal symptoms are relieved if the drug is taken. The withdrawal states are further classified as:
   (*i*) Uncomplicated
   (*ii*) With convulsions
   (*iii*) With delirium

3. *Dependence Syndrome* is a condition, which generally develops on regular and continued use of a drug. When a person becomes dependent on a particular drug, he has a compulsive need to obtain and use the drug, each time requiring a higher dose of the drug to get the same effect because of development of tolerance. In case the drug is not available or available in lesser quantity, withdrawal symptoms develop. This behaviour is detrimental to personal and social life of the individual. According to the International Classification of Diseases (ICD-10), the dependence syndrome is a cluster of physiological, behavioural, cognitive phenomena in which the use of a substance or a class of substances taken on a much higher priority for a given individual than other behaviours that once had greater value. The dependence syndrome is characterised by a strong, often over-powering desire to take a psychoactive drug (may or may not have been prescribed medically). The definite diagnosis of dependence should usually be made only if at least three of the following have been experienced or exhibited at some time during the previous year:
   (*i*) A strong *desire* or *sense of compulsion* to take the substance.
   (*ii*) Difficulties in controlling substance-taking behaviour in terms of its onset, termination or levels of use.
   (*iii*) A physiological *withdrawal state* occurs when the substance use has ceased or reduced.
   (*iv*) Evidence of *tolerance* is seen in dependent users. In order to achieve the effects of the drug obtained with low doses initially, the person has to take the higher doses.
   (*v*) Progressive *neglect* of alternative pleasures or interests because of substance use occurs and the individual spends more time in procuring and using the psychoactive drug.

(*vi*) There is persistence with the substance despite of harmful effects caused by the drug on various body organs.

4. *Harmful Use* of a drug can be classified as follows:
   (*i*) Continuous drug use, despite the awareness of harmful medical and/ or social effect of the drug being used.
   (*ii*) A pattern of physically hazardous use of drug (i.e., driving during intoxication)

The diagnosis of 'Harmful use requires that the actual damage is caused to the physical or the mental health of the user. If the patient qualifies for the diagnosis of dependence syndrome then harmful use is not diagnosed. ICD-10 includes other syndromes associated with psychoactive substance use such as psychotic disorder, amnesic syndrome, and residual and late-onset psychotic disorder.

The major dependence producing drugs are as shown in Table 1.

Table 1: Psychoactive Substances

1. Alcohol
2. Opioids i.e., opium, heroin
3. Cannabinoids i.e., Cannabis, Marihuana
4. Cocaine
5. Amphetamines and other sympathomimetic drugs
6. Hallucinogens i.e., Lysergic acid diethylamide (LSD), Phencyclidine (PCP)
7. Sedatives and hypnotics i.e., barbiturates
8. Inhalants i.e., volatile solvents
9. Nicotine, and
10. Other stimulants i.e., Caffeine

Aetiology of Substance Use Disorder

It is a complex interplay of multiple factors that leads to this disorder. Biological, psychological and social factors have been implicated described in Table 2.

Table 2: Aetiological Factors in Substance Use Disorders

1. **Biological Factors**
   (*i*) Genetic vulnerability
   (*ii*) Co-morbid psychiatric disorder or personality disorder
   (*iii*) Co-morbid medical disorders
   (*iv*) Reinforcing effects of the drugs
   (*v*) Withdrawal effects and craving
   (*vi*) Biochemical factors

Contd...

2. Psychological Factors

(*i*) Curiosity, need for novelty seeking
(*ii*) General rebelliousness and social non-conformity
(*iii*) Early initiation of alcohol and tobacco
(*iv*) Poor impulse control
(*v*) Sensation-seeking (high)
(*vi*) Low self-esteem (anomie)
(*vii*) Concerns regarding personal autonomy
(*viii*) Poor stress management skills
(*ix*) Childhood trauma or loss
(*x*) Relief from fatigue or boredom
(*xi*) Escape from reality
(*xii*) Lack of interest in conventional goals
(*xiii*) Psychological distress

3. Social Factors

(*i*) Peer pressure (often more important than the parental factor).
(*ii*) Role model: imitating the ego-ideal.

Complications

Alcohol dependence has its effect not only on the consumer's body but also on the social and personal life of the individual leading to several complications both medical and social. These are as follows:

1. Acute Intoxication

After initial excitation for a brief period, there is generalised central nervous system depression with alcohol use. With increasing intoxication, there is increased reaction time, slowed thinking and distractibility and poor motor control. Later, dysarthria, ataxia and incoordination occur. There is progressive loss of self control with frank disinhibited behaviour. The duration of intoxication depends on the amount and the rapidity with which alcohol is ingested. The signs of intoxication are obvious with blood alcohol levels of 150–200 mg%. Increasing drowsiness followed by coma and respiratory depression develop at the blood level of 300–450 mg%. When the blood levels reach between 400–800 mg%, death is likely to occur (Table 3).

Table 3: Blood Alcohol Levels and Behaviour

Blood alcohol concentration	*Behavioural correlated*	
25–100 mg%	Excitement	
80 mg%	Is the legal limit for driving in UK while in India this limit is 30 mg%	
100–200 mg%	Serious intoxication, slurred speech, incoordination, nystagmus.	
300–350 mg%	Hypothermia, dysarthria, cold sweats	
350–400 mg%	Coma	
>400 mg%	Death may occur	
Urinary alcohol concentration	*Diagnostic use*	*Equivalent blood alcohol concentration*
>120 mg%	Suggestive	80 mg%
>200 mg%	Diagnostic	150 mg%

Sometimes a small dose of alcohol may produce active intoxication in some persons. It is an idiosyncratic reaction to alcohol and not related to blood levels. This condition is known as *pathological intoxication*. Sometimes an individual develops complete amnesia of the acute intoxication known as *blackouts*.

2. Withdrawal Syndrome

In a dependent alcoholic when the blood levels decrease withdrawal symptoms appear. The most common withdrawal syndrome is a hangover on the next morning. Mild tremors, nausea, vomiting, weakness, irritability, insomnia and anxiety are other common withdrawal symptoms. Sometimes withdrawal syndrome is more severe characterised by one of the following three: alcoholic hallucinosis, alcoholic seizures, and delirium tremens.

(*i*) *Alcoholic hallucinosis*: During partial or complete abstinence, a dependent alcoholic experiences hallucinations (usually auditory). The hallucinations are generally accusatory or threatening in nature. It occurs in about 2% of the cases. These hallucinations persist after the withdrawal syndrome is over, and classically occur in clear consciousness. Usually recovery occurs within one month and the duration is very rarely more than six months.

(*ii*) *Alcoholic seizures* (rum fits): Generalised tonic clonic seizures occur in about 10% of alcohol dependent patients, usually 12–48 hours after a heavy bout of drinking. Often, these patients have been drinking alcohol in large amounts on a regular basis for many years. Multiple seizures (2–6 at one time) are more common than single seizures. Sometimes status epilepticus may be precipitated. In about 30% of the cases, delirium tremens follows.

(*iii*) *Delirium tremens* (DT): It is the most severe alcohol withdrawal syndrome. It occurs usually within 2–4 days of complete or significant abstinence from heavy alcohol drinking in about 5% of patients, as compared to acute tremulousness which occurs in about 34% of the cases. The course of delirium tremens is short and recovery generally occurs within 7 days. This is an acute organic brain syndrome with the following characteristic clinical features:

(a) Clouding of consciousness with disorientation in time and place.

(b) Poor attention span and distractibility.

(c) Visual (and also auditory) hallucinations and illusions, which are often vivid and very frightening. Tactile hallucinations of insects crawling over the body may occur.

(d) Marked autonomic disturbance with tachycardia, fever, sweating, hypertension and papillary dilatation are the usual features encountered.

(e) Psychomotor agitation and ataxia may be present.

(f) Insomnia, with a reversal of sleep-wake pattern can be seen.

(g) Dehydration with electrolyte imbalance is present.

Death may occur due to cardiovascular collapse, infection, hyperthermia or self-inflicted injury. At times, intercurrent medical illness like pneumonia, fractures' liver disease and pulmonary tuberculosis may complicate the clinical picture.

3. Neuropsychiatric Complications of Chronic Alcohol Use

(*i*) *Wernicke's encephalopathy*: This is an acute reaction occurring in response to severe deficiency of thiamine, commonest cause being chronic alcohol intake. The onset of this disorder characteristically occurs after a period of persistent vomiting. The important clinical features are:

(a) Ocular signs: Course nystagmus and ophthalmoplegia, with bilateral external rectus muscle paralysis occur early. In addition, papillary irregularities, retinal haemorrhages and papilledema can occur, causing an impairment of vision.

(b) Higher mental function disturbance: Disorientation, confusion, recent memory disturbances, poor attention span and distractibility are the common impairments of higher mental functions. Other early symptoms seen are apathy and ataxia. Peripheral neuropathy and serious malnutrition are generally co-existent. Neuropathological findings show neuronal degeneration and haemorrhage in thalamus, hypothalamus, mammillary bodies and midbrain.

(*ii*) *Korsakoff's psychosis*: Korsakoff's psychosis often follows Wernicke's encephalopathy and together they are referred as the Wernicke-Korsakoff Syndrome. Clinical manifestations of Korsakoff's psychosis include an organic amnestic syndrome, characterised by gross memory disturbances with confabulation and impaired insight. Neuropathological findings show widespread lesion and the most consistent changes are seen in bilateral dorsomedial nuclei of the thalamus and mammillary bodies. The changes are also seen in periventricular and periequiductal grey matter, cerebellum and parts of brainstem. The cause is usually severe, untreated thiamine deficiency secondary to chronic alcohol use.

(*iii*) *Marchiafava-Bignami disease*: This disorder is probably caused by alcohol-related deficiency with pathological changes in the form of widespread demyelination of corpus callosum, optic tract and cerebellar peduncles. Clinical manifestations include disorientation, epilepsy, ataxia, dysarthria, hallucinations, spastic limb paralysis, and deterioration of personality and intellectual functions. This is a rare disorder.

(*iv*) *Other complications of alcohol use include*

(a) Alcoholic dementia

(b) Cerebellar degeneration

(c) Peripheral neuropathy

(d) Central Pontine myelinosis

(e) Optic atrophy (particularly with methyl alcohol)

Treatment

Following steps should be considered before starting treatment for alcohol dependence:

- Look for the possibility of a physical disorder.
- Look for the possibility of a psychiatric diagnosis.
- Patient's motivation to undergo treatment should be assessed.
- Assessment of the social support the patient enjoys should be done.
- Personality characteristics of the patient should be assessed in details.
- Level of current and past occupational and socials functioning should be assessed.

The treatment comprises two broad components i.e., detoxification and treatment of alcohol dependence.

1. Detoxification

Treatment of alcohol withdrawal symptoms is detoxification. Symptoms that are produced due to non-availability or lesser availability of alcohol to a dependent alcohol user are withdrawal symptoms and the constellation of symptoms form withdrawal syndrome. The usual duration of uncomplicated withdrawal syndrome is 7–14 days. The aim of detoxification is the symptomatic management of the emergent withdrawal symptoms.

Benzodiazepines are the drugs of choice for symptomatic detoxification. Chlordiazepoxide is generally used in the dosage range of 80–200 mg per day in three to four divided doses. Another drug used is diazepam, its dose being 40–80 mg per day in divided doses. The typical dose of chlordiazepoxide in moderate alcohol dependence should be as follows:

— 1st day 20 mg four times
— 2nd day 15 mg four times
— 3rd day 10 mg four times
— 4th day 5 mg four times
— 5th day 5 mg two times and then stop.

In severe dependence, higher doses are required for longer period (up to 10 days)

In some countries drugs like chlormethiazole (1–2 g/day) and carbamazepine (600–1600 mg/day) are also used for detoxification.

These drugs follow a standard dose regime with dosage steadily decreasing and stopping on 10th day.

Nutritional deficiency is generally associated with alcohol dependence therefore, vitamins should be given.

2. **Treatment of Alcohol Dependence:** After the detoxification process is over further management is done with the aim to prevent the patient from restarting alcohol consumption. Several methods are available to choose that suit best to the patient. Some of these important methods are as follows:
   (*i*) Behaviour therapy
   (*ii*) Psychotherapy
   (*iii*) Group therapy
   (*iv*) Deterrent therapy
   (*v*) Other medications
   (*vi*) Psychosocial rehabilitation

Opioid use Disorder

The natural alkaloids of opium and their synthetic preparations are highly dependence producing drugs (Table 4). The dried exude obtained from unripe seed capsules of *Papaver somniferum* is a highly dependence producing and is abused for centuries. In the last few decades, the use of opioids has tremendously increased all over the world. India being the transit point for illicit drug trade between golden triangle (Burma-Laos-Thailand) and golden crescent (Iran-Afghanistan-Pakistan) is among the worst effected countries. Addition of heroin to Indian streets, some two decades ago, has caused devastating effects on the adolescents and the youth of the country.

Table 4: Opioid Derivatives

1. **Natural Alkaloids of Opium**
   (*i*) Morphine
   (*ii*) Codeine
   (*iii*) Thebaine
   (*iv*) Noscapine
   (*v*) Papaverine
2. **Synthetic Compounds**
   (*i*) Heroin
   (*ii*) Nalorphine
   (*iii*) Hydromorphine
   (*iv*) Methadone
   (*v*) Dextropropoxyphene
   (*vi*) Meperidine (Pethidine)
   (*vii*) Cyclazocine
   (*viii*) Levallorphan
   (*ix*) Diphenoxylate

Morphine and heroin are the most potent dependence producing derivatives; they bind to *mu* opioid receptors. The other opioid receptors are *kappa* (for pentazocine), *delta* (for a type of encephalin), *sigma* (for phencyclidine), *epsilon* and *lambda.* Heroin (diacetyl-morphine) is two times more potent than morphine in injectable form. Apart from the parenteral mode of administration, heroin can also be smoked or chased (*chasing the dragon*), often in impure form (popularly known as *smack* or *brown sugar* in India. Due to its higher addictive potential than morphine, heroin causes dependence after a short period of exposure. Tolerance to heroin occurs rapidly and can be increased up to more than 100 times the first dose needed to produce an effect.

Acute Intoxication

Acute intoxication due to opioids is characterized by apathy, bradycardia, hypotension, respiratory depression, subnormal core body temperature and pin-point pupils. Subsequently, reflexes become delayed, pulse becomes thready and coma may occur in case of a large overdose. In severe intoxication, mydriasis may occur due to hypoxia.

Withdrawal Syndrome

Withdrawal symptoms generally appear within 12–24 hours, reach their peak within 24–72 hours and subside within 7–10 days of the last dose of opioids. The characteristic symptoms include lacrimation, rhinorrhoea, papillary dilatation, sweating, diarrhoea, yawning, tachycardia, mild hypertension, insomnia, raised body temperature, muscle cramps, generalised bodyache, severe anxiety, piloerectio, nausea, vomiting and anorexia. There can be marked individual differences in manifestation of withdrawal symptoms. Withdrawal syndrome with heroin is more severe as compared to the one seen with morphine.

Complications

Chronic opioid use may lead to several complications. The important ones are listed here. One may experience one or more of the following complications:

(*i*) *Complications due to illicit use* are quite common. Illicit drug is usually contaminated with some toxic additives. Complications generally seen are parkinsonism, degeneration of globus pallidus, peripheral neuropathy, amblyopia, transverse myelitis.

(*ii*) *Complications due to intravenous use* are quite common if the needles are exchanged. The user gets exposed to the risk of various infections. The complications generally seen are AIDS, skin infections, thrombophlebitis, pulmonary embolism, septicemia, viral hepatitis, tetanus, endocarditis.

(*iii*) *Drug pedalling and criminal activities* may become a part of chronic users' life leading to various social and legal complications. Production, manufacture, import, export, sale, purchase and even use of opioids is illegal and is liable to be penalised with severest of punishment (under Narcotic Drugs and Psychotropic Substances Acts 1985 in India). Chronic use also leads to financial and occupational difficulties leading to marital dicord and other social problems.

Treatment

Before starting treatment correct diagnosis should be established on the basis of detailed history of drug intake, the examination and laboratory tests. These tests provide an evidence of drug use. These tests include, Naloxone challenge test to precipitate withdrawal symptoms and urinary opioid testing with radioimmunoassay (RIA), free radical assay technique (FRAT), thin layer chromatography (TLC), gas liquid chromatography (GLC), high pressure liquid chromatography (HPLC) or enzyme-multiplied immunoassay technique (EMIT). After confirming the diagnosis, the treatment can proceed on three lines:

1. Treatment of overdose
2. Detoxification
3. Maintenance therapy

1. **Treatment of Opioid Overdose:** Narcotic antagonists (i.e., naloxone, naltrexone) are generally used to treat opioid overdose. Intravenous injection of 2 mg naloxone followed by repeated injection in 5–10 minutes, causes reversal of overdose. Since naloxone has a short half-life, repeated doses are required every 1–2 hour. General supportive care should be provided along with.
2. **Detoxification:** is the process in which a opioid dependent person is "freed" from opioids. This is usually done by abruptly stopping the opioid, and managing the emergent withdrawal symptoms. Detoxification of opioid dependents is highly successful, relatively cheaper, applicable on a very large scale, associated virtually with no morbidity or

mortality and acceptable to almost all patients. Following methods are available for the management of the withdrawal symptoms:

(*i*) *Use of substitution* drugs such as methadone is quite common though this drug is not available in India, to control withdrawal symptoms. Methadone is relatively less addictive, has longer half-life, decreases possible criminal behaviour and has much milder withdrawal symptoms when stopped. The aim is to gradually taper the patient from methadone. Relapse rates are quite high when methadone is stopped. It is also argued that one dependence is replaced by another.

(*ii*) *Clonidine* is an α_2 agonist that acts by inhibiting norepinephrine release at the presynaptic α_2 receptor. The usual dose of clonidine is 0.3–1.2 mg per day, and it is tapered off within 10–14 days. It is started after stopping the opioid. Clonidine causes excessive sedation and postural hypertension, therefore the treatment is ideally started in the indoor setting. Regular blood pressure monitoring should be done.

(*iii*) *Naltrexone* in combination with clonidine is used for the treatment of opioid dependence. Naltrexone is a orally available narcotic antagonist blocking the action of opioids in a dependent person and thus causing withdrawal symptoms. These withdrawal symptoms are managed with addition of clonidine for a period of 10–14 days and then stopping it. Subsequently the patient continues on naltrexone alone. Now if the person takes an opioid, there are no pleasurable experiences as the opioid receptors are blocked by naltrexone. Therefore, this method can be regarded as the combination of detoxification and maintenance treatment. The usual dose of naltrexone is 100 mg orally, administered every alternate day.

(*iv*) *Other drugs* are also used as detoxification agents. These are:

(a) LAAM (levo-alpha-acetyl-methadol) is not in wide use for opioid dependence and its use as a long-term treatment agent began in late 1960s. Its analgesic effect and delayed action and longer duration of action were first noted in animals in 1948. With subcutaneous injection of 10–30 mg LAAM analgesic effect is produced after 4–6 hours lasting for 48–72 hours. Intravenous administration also produces similar results, while with oral administration effect appears much quicker (within 1–2 hours) and persists for 72 hours.

(b) Propoxyphene

(c) Diphenoxylate

(d) Buprenorphine is a long-acting partial *mu*-agonist and is commonly used for both detoxification and maintenance treatment.

(e) Lofexidine is a α_2 agonist like clonidine.

3. **Maintenance Therapy:** After the patient is detoxified, the next step is to put the patient on maintenance therapy. The patient is maintained on one of the following regimes:

(i) Methadone maintenance (Agonist substitution therapy)
(ii) Opioid antagonists
(iii) Other methods
(iv) Psychosocial rehabilitation

Cannabis use Disorder

Cannabis is derived from hemp plant *Cannabis sativa* and different parts of the plant yield the products popularly known as *grass, hash* or *hashish, marihuana, charas, bhang* etc. The plant carries more than 400 identifiable chemicals of which about 50 are cannabinoids, the most active being:

Table 5: Cannabis Preparation and THC Content

Cannabis preparation	*Part of the plant from where it is obtained*	*THC content (%)*	*Potency as compared to Bhang*
Hashish/Charas	Resinous exudates from the flowering tops of cultivated plants	8–14%	10
Ganja	Small leaves and brackets of inflorescence of highly cultivated plants	1–2%	2
Bhang	Dried leaves, flowering shoots and cut tops of uncultivated plants	1%	1
Hash oil	Lipid soluble plant extract	15–40%	25

Cocaine use Disorder

Cocaine is an alkaloid derived from the coca bush, *Erythroxylum coca*, found in Bolivia and Peru. It was isolated by Albert Neimann in 1860 and was used by Karl Koller, a friend of Sigmund Freud in 1884 as the first effective local anaesthetic agent. In the last few decades, cocaine has gained popularity as a street drug, *crack* and can be administered in the body orally, intranasally, by smoking or parenterally, depending upon the preparation available. The commonest forms used are cocaine hydrochloride and free base alkaloid. Both intravenous use and free base inhalation produce a 'rush' of pleasurable sensations.

Cocaine is central stimulant, which inhibits the reuptake of dopamine, along with the reuptake of norepinephrine and serotonin. In animals, cocaine is the most powerful reinforcer of the drug-taking behaviour. Cocaine is, sometimes, used in combination with opiates like heroin ('speed ball') or at times amphetamines. Cocaine has started making its presence felt in India's bigger cities.

Acute effects are due to central stimulant and sympathomimetic effects as euphoria, confidence, increased energy, increased heart rate and blood pressure, dilated pupils, constriction of peripheral blood vessels and rise in body temperature and metabolic rate. In an intoxicated state, the user presents with papillary dilatation, tachycardia, hypertension, sweating and nausea or vomiting. A hypomanic picture with increased psychomotor activity, grandiosity, elation of mood, hypervigilance and increased speech output may be present. Later, judgment is impaired and there is impairment of social and occupational; functioning.

In various trials it has been found that the effects of cocaine on heart, blood pressure respiratory rate, and mood increase as the dose is raised from 4 mg to 30 mg major effects are observed with the dose 16 mg and above. In higher doses cocaine can cause depression of the medullary centres and death from cardiac, and more often respiratory arrest. Because cocaine causes increased energy and confidence and can produce irritability and paranoia, it may lead to physical aggression and crime.

Chronic effects are generally not marked if cocaine is used for 2–3 times a week for recreational purposes. Taken daily in fairly large amounts, it can disrupt eating and sleeping habits, produce minor psychological disturbances including irritability and difficulty in concentration, and create a serious psychological dependence. Perceptual disturbances (especially pseudohallucinations), paranoid thinking and rarely psychosis also occur in chronic users of cocaine. A runny and clogged nose is common to be seen and that can be treated with nasal decongestant sprays. Less often, the nose can become inflamed, swollen or ulcerated. Rarely, perforated nasal septum is also reported.

Dependence potential varies with the route of administration. Intravenous cocaine use is most powerful drug-reinforcer.

Withdrawal Syndrome

Cocaine does not produce physical dependence in the sense that alcohol and heroin does but sometimes mild withdrawal symptoms such as anxiety and depression arise. Physical dependence is very mild if it is there but the psychological dependence is very strong.

Table 6: Complications due to Cocaine use

The complications of chronic cocaine withdrawal include:

1. Acute anxiety reaction
2. Uncontrolled compulsive behaviour
3. Psychotic episode (with persecutory delusions and acytile and other hallucinations)
4. Delirium and delusional disorder
5. Seizures (especially with high dose)
6. Respiratory depression
7. Cardiac arrhythmias
8. Coronary artery occlusion
9. Myocardial infarction
10. Lung damage
11. Gastrointestinal necrosis
12. Fetal anoxia
13. Nasal septum perforation.

Treatment

Inhalants

Volatile solvents have the capacity to intoxicate those who deliberately inhale them. Hundreds of intoxicating volatile products are available in the homes as well as market place much within the reach of adolescents and young people. Use of inhalants is found more common among school going adolescents and inhalation of these agents is a peer-originated and peer-perpetuated activity. There is a wide range of inhalants commonly abused:

(*i*) Airplane glue
(*ii*) Fingernail polish remover
(*iii*) Gasoline
(*iv*) Paint thinner
(*v*) Liquid shoe polish
(*vi*) Plastic cement
(*vii*) Cleaning fluid
(*viii*) Wax strippers
(*ix*) Petrol
(*x*) Kerosene oil

Causes

(*i*) Indulgence in inhalant use behaviour is generally caused by unsuccessful and unrewarding school experiences.

(*ii*) Personality deficiencies are reported to be important predisposing factors in confirmed inhalant abusers.

(*iii*) Youngsters overwhelmed with anxiety, depression, or both; boarderline or over schizophrenia; and those with character disorders employ inhalants in effect at self-treatment for their intrapsychic and interpersonal distress.

(*iv*) Social disorganisation within the community also contribute to this practice.

Consequences

(*i*) Psychiatric features of acute intoxication resemble those of alcohol intoxication except for its brief duration. The period of relative stimulation and of disinhibited behaviour is also similar to that of alcohol. This can result in accidental injury or death and the releasing of aggressive impulses against one's self or others. Cognitive deficit may occur with extensive and prolonged inhalant abuse. Psychological maturation gets arrested and aberrant behaviour is not uncommon. School-dropout is a usual consequence of inhalant abuse.

(*ii*) Physical symptoms may occur due to the toxic effects of the inhalants. Some of the volatile solvents are the known poisons. Carbon tetrachloride is so toxic that it has been removed from the commercial trade, and benzene's use is limited for the same reason.

Hexane and leaded gasoline can cause of a serious polyneuritis, and the latter is capable of producing encephalopathy. Toluene is involved in dysfunction of kidney, nervous system and bone marrow. Metallic spray paints and other aerosols may have dangers caused by secondary ingredients rather than by solvent themselves. Sudden sniffing deaths have been reported. Ventricular fibrillation and other arrhythmias occur.

Benzodiazepines and other Sedative-Hypnotic use Disorder

These drugs are generally used in the treatment of anxiety and insomnia. Currently, these are the most often prescribed drugs. Chlordiazepoxide was discovered by Sternbach in 1957 and since then benzodiazepines have replaced other sedative-hypnotic drugs. Benzodiazepines produce their effects by acting on benzodiazepine receptors (GABA-Benzodiazeine receptor complex), thereby indirectly increasing the action of GABA, the chief inhibitory neurotransmitter in the human brain.

Benzodiazepine (or other sedative-hypnotic) use disorder can either be iatrogenic or originating with illicit drug use. Dependence, both psychological and physical, can occur and tolerance is usually moderate.

Intoxication and Complications

Withdrawal Syndrome

Treatment

Tobacco

Tobacco is a plant product derived mainly from *Nicotiana tobaccum* and *Nicotiana rustica* grown all over the world and used by people in almost all countries. Tobacco plant is a native of Americas and was brought by the European navigators on their discovery of Americas in 1492. Tobacco leaves contain an active alkaloid, which is highly toxic and develops resistance to its own action and hence highly addictive. Tobacco is used in smokeless form and it is smoked as cigarette, beedi etc. Withdrawal symptoms include, restlessness, nervousness, anxiety, insomnia, increased weight. For tobacco-cessation, pharmacological and behavioural techniqiues are used. First line treatment includes Bupropion (150 mg per day for 3 days and then 150 mg twice a day for 6–7 weeks). Nicotine replacement therapy in the form of nicotine patches, nicotine spray, nicotine lozenges is available. Secondline treatment includes clonidine and amitriptyline. Behaviour therapy and counselling are used along with pharmacotherapy or independently.

8

Anxiety Disorders

Introduction

Anxiety is a common emotion that we experience almost daily. One could say one is living in an 'age of anxiety.' Appearing for an examination, meeting new people, getting a parking ticket, an encounter with the boss, all can lead to anxiety. Anxiety is characterised by a diffuse vague sense of unpleasantness, apprehension, accompanied with autonomic symptoms like palpitation, choking, tightness in the abdomen, muscular tension and restlessness. There may be different constellations of physiological responses to the anxiety in different people.

Fear is also an emotion we are all familiar with. When we see danger we either fight it or run away from it. Species that have learnt to fight well or flee well, have survived. The typical physiological response to fear prepares the body for such a fight or flight response. The heart beats faster, the muscles become more tensed, the breathing becomes faster, pupils dilate so that the eyes can see better, the body becomes "primed" for a quicker response. Fear also has an adaptive function. So fear is an important emotion, particularly for the preservation of life.

Anxiety has an adaptive function too. Anxiety is an alerting signal. It warns us of impending danger, and primes us to deal with the danger. Anxiety is similar to fear. Fear is in response to external danger or threat, as in fear of a dog, or fear of a man with a gun, whereas anxiety is in response to internally perceived danger or threat. In the sense the both fear and anxiety are alerting in nature, and prepare us to cope with danger, both fear, and anxiety are adaptive responses. The extent to which an individual is able to cope with a certain anxiety depends on the nature of the threat and the person's coping mechanisms, and psychological strengths.

The experience of anxiety has two components. One is the awareness of physiological sensations like increased heart rate, sweating, and the other is the awareness of being nervous. Anxiety may also be associated with a feeling of shame—that "others will recognise that I am afraid or anxious." In a state of anxiety a person's cognitive or intellectual skills may also be affected—there may be a tendency to catastrophic thinking, a person may misinterpret benign arousal sensations, or may have faulty or selective recall of stressful situations.

Anxiety as An Illness

In recent times, anxiety is being recognised more and more as an illness. In the Diagnostic and Statistical Manual of Mental Disorders, there are changes in the conceptualisation of anxiety in each new edition. There has been a moving away from the psychodynamic concept of a *neurosis* to a classification based on clinically recognisable criteria.

For the clinician it is important to distinguish between normal and pathological anxiety. Pathological anxiety is clearly out of proportion to the degree of stress, the duration, the severity and the frequency of symptoms are far in excess of what would be considered as a normal response to a danger or threat.

Etiology and Pathology of Anxiety

Psychoanalytically, anxiety is the result of poor conflict resolution or repression of an unacceptable drive or impulse. Impulse anxiety, separation anxiety, castration anxiety and superego anxiety are hypothesised to develop at various stages of growth.

The behavioural theory of anxiety proposes that anxiety is a conditioned response to environmental stress. Anxiety also may be the result of catastrophic misinterpretation of benign arousal sensations. The behaviourists have been able to come up with the most effective non-pharmacological treatments for anxiety disorders.

The biological basis of understanding anxiety disorders is by studying neuroanatomical structures involved in anxiety, and the neurochemical correlates of anxiety. The neuroanatomical structures involved in the processing of anxiety are the amygdala, the thalamus and other structures in the limbic system comprising what is called the fear network. The neurochemical mediators of anxiety are through Gamma Amino Butyric Acid (GABA) mediated inhibitory pathways, Glutamate mediated excitatory pathways and some neuromodulation by serotonergic pathways. Most benzodiazepines, which are excellent antianxiety drugs, work by affecting the GABAergic neurons.

DSM-IV Classification of Anxiety Disorders

1. Panic disorder with and without Agoraphobia
2. Social phobia and specific phobias
3. Obsessive compulsive disorder
4. Posttraumatic stress disorder
5. Acute stress disorder
6. Generalised anxiety disorder
7. Anxiety disorder due to a general medical condition
8. Substance induced anxiety disorder
9. Anxiety disorder not otherwise specified including mixed anxiety depressive disorder.

Panic Disorder with or without Agoraphobia

Panic is a paroxysmal occurrence of intense fear or discomfort which is either completely unexpected, or is cued by a situation. The panic attacks are recurrent, spontaneous, and last more than a month. The patient shows persistent concern about more attacks or their implications. Women are twice as more likely to be affected. Women have more agoraphobia, men more likely to have panic alone, and are more likely to self medicate with alcohol or other prescription drugs to cut off the panic attack.

Lifetime prevalence of panic attacks is 1% to 4%. But in populations who actively seek treatment, the percentage may be as high as 10% to 60%. Onset of the illness is in late teenage, or 25 to 30 yrs. Almost 90% of panic disorder patients have co-morbid illness like other anxiety disorder, mood disorders, substance use disorder or personality disorders. If untreated, there is a high suicide risk. Patients with panic disorder have poor quality of life due to the utterly incapacitating nature of their illness. Most of them have impaired social and occupational functioning. Panic disorder patients are the most frequent users of the emergency services, and often undergo long repeated costly investigations to rule out what they imagine as a serious heart ailment, or a sudden death situation. The illness follows a chronic non-remitting course, with only 30% patients showing symptom remission, almost 40–50% have improved but still have significant symptoms, and nearly 20–30% remain unimproved.

A detailed, open-ended interview is the most important tool for establishing a diagnosis of panic disorder. After ruling out general medical disorders which could mimic a panic attack, repeated rounds of investigations should be avoided.

Treatment is with tricyclic antidepressants like imipramine and amitryptiline, benzodiazepines like diazepam, alprazolam, lorazepam etc and most importantly SSRIs like Fluoxetine, sertraline, paroxetine, etc. The possibility of dependence on benzodiazepines is very high, so the drug use should be closely monitored.

Cognitive behaviour therapy with a lot of psychoeducation, relaxation training, breathing retraining is most effective in patients with panic disorder.

Social Phobia and Specific Phobias

Phobias are most common anxiety disorder; in fact phobias are the most common of all mental illnesses. Most of them go unrecognised or untreated because the person suffering from them does not feel the need to seek treatment.

Social phobia is a marked and persistent fear of one or more social or performance situations in which a person is exposed to unfamiliar people or to possible scrutiny of others. Meeting new people, interacting with them socially, attending parties, meetings, speaking formally, eating or writing in front of others, dealing with people in authority, are all situations where a person with social phobia will be greatly stressed. Although social phobias are very common, it is only recently that they have been recognised as a separate diagnostic entity.

Specific phobias are a marked and persistent fear of a specific object or situation. The patient also recognises it as an unreasonable response, and avoids it or endures it with great distress. Phobias may be for animals or objects, for natural environment like water bodies, mountains, sea etc. for situations like closed spaces, dark spaces, lifts etc. or for blood or injury.

Lifetime prevalence of phobias is 11–15%. Women are affected twice as much as men. Social phobias are found in 2% of the population. Many phobias are associated with co-morbid illnesses like major depressive disorder or other anxiety disorders.

In treating phobias, the most effective pharmacological agents are the SSRIs and benzodiazepines. But behaviour therapy and cognitive behaviour therapy are the most effective treatment options. The most important goal of therapy should be to reduce fear and phobic avoidance to manageable levels.

Obsessive Compulsive Disorder

Obsessive compulsive disorders were once considered rare, but obsessive compulsive disorder is the fourth most common mental illness, and shows a lifetime prevalence of 2 to 4% in the population. More women than men have OCD, but in men there are more co-morbid disorders. In childhood OCD, males are affected twice as much as females.

Obsessions are intrusive, recurrent, unwanted ideas, thoughts or impulses that are difficult to dismiss, which give rise to anxiety. The patient may feel guilty about the obsessive thoughts especially if the thoughts are religious or sexual in nature. In some patients, the obsessive thoughts remain constant, whereas in others they change in content but are associated with the same degree of distress. Compulsions are repetitive behaviours, which may be observable or only mental, that are intended to reduce the anxiety of obsessions. Most illnesses start in early childhood, but become significantly distressing to the patient to warrant treatment only around puberty or early adulthood. Familial OCD tends to have earlier onset. The obsessive thoughts can be described as aggressive obsessions, contamination obsessions; sexual, religious, or somatic obsessions. Compulsive acts may be either checking, ordering, rearranging, following a strict order or symmetry, repeating rituals, religious rituals, or mental rituals. Most patients have both obsessions and compulsions.

The illness follows a chronic non-remitting course, with waxing and waning symptoms. At any given point of time only 20% patients are symptom free, 40–30% have recovery with subclinical symptoms, and about 30–35% have recovery but with significant symptoms. Co-morbid disorders are frequent, of which depression is the most common. Tic disorders, delusional or schizotypal personality disorders also coexist.

There is a strong genetic linkage in Obsessive Compulsive Disorders. 63% monozygotic twins are concordant for OCD. There is strong clinical suggestion that OCD is one of many Obsessive Compulsive spectrum disorders like Asperger's syndrome, hypochondriasis, kleptomania, tic disorder, Tourette's disorder, obsessive personality disorder, body dysmorphic disorder etc. These may be a different group of disorders with imbalance in the serotonergic dopaminergic systems, and involving disturbances in orbitofrontal cortex, the basal ganglia and the thalamus.

Treatment is with SSRIs and benzodiazepines. Often augmentation with other drugs like clonazepam, lithium, risperidone etc. helps to fine tune the treatment response.

Behaviour therapy, especially cognitive behaviour therapy is found to be as effective as pharmacotherapy alone. A judicious combination of CBT and pharmacotherapy is effective in most patients. In some extreme cases neurosurgical interventions have also been used.

Post Traumatic Stress Disorder

PTSD is found in persons who are exposed to severely traumatic events. Such trauma may be war, natural calamities, and rape, riots, auto accidents, building fires, bombings etc. The person experiences flashbacks—re-experiencing the trauma through dreams or waking thoughts, and shows a persistent avoidance to reminders of the trauma, and a state of hyperarousal. The significant distress in social or occupational functioning lasts more than a month.

PTSD affects females twice as much as males. There is a 8% lifetime prevalence in exposed populations. Treatment consists of supportive therapy and crisis management, encouragement to talk, cognitive behaviour therapy and Group therapy. Pharmacological treatment includes SSRIs especially sertraline, and paroxetine, and benzodiazepines.

Acute Stress Disorder

This is similar to PTSD, but occurs in response to and extraordinary mental or physical stressor. Symptoms appear within a few hours to a month of witnessing the stressor and include social withdrawal, aggression, disorientation, grief, hopelessness and despair. Symptoms last a short while, from a few days to four weeks. Treatment is similar to PTSD.

Generalised Anxiety Disorder

These are the patients most likely to come to a general practitioner, and most of them are referred to the psychiatrist. Patients present with excessive tension or apprehension about everyday problems. Duration of symptoms is for at least six months. Patients present with symptoms of panic, increased muscle tension, non-specific sleep disturbances, reduced concentration, irritability, 'mind going blank' etc.

To make a diagnosis, one must rule out medical conditions which may mimic anxiety, personality disorders or somatoform disorders.

Treatment is with newer antidepressants SSRIs, Benzodiazepines, venlafaxine or buspirone. These drugs are used in accordance with the severity of the illness. Benzodiazepines are generally used for a shorter period because of their addictive properties. Psychotherapeutic intervention is quite helpful in these clinical conditions. Insight oriented psychotherapy and Cognitive Behaviour Therapy (CBT) is also helpful.

9

Somatoform Disorders

The ***essential features*** of this group of disorders are physical symptoms suggesting physical disorder (hence somatoform) for which there are no demonstrable organic findings or known physiological mechanisms and for which there is positive evidence, or a strong presumption that the symptoms are linked to psychological factors or conflicts.

The common somatoform disorders are:

I. SOMATISATION DISORDER (BRIQUET'S SYNDROME)

It is primarily characterised by the presence of recurrent and multiple somatic complaints of several years duration for which medical attention has been sought but which was apparently not due to any physical disorder.

Symptoms usually begin in the teen years or rarely in the twenties.

CLINICAL FEATURES

Symptoms from at least two systems must be present continuously for at least two years.

Gastrointestinal symptoms

Vomiting (other than during pregnancy)—abdominal pain (other than when menstruating), nausea, bloating.

Pain symptoms

Pain in extremities, back pain.

Cardiopulmonary symptoms

Shortness of breath when not exerting oneself, palpitations. Chest pain, dizziness.

Conversion or Pseudoneurological symptoms

Amnesia—difficulty swallowing, loss of voice, fainting or loss of consciousness, seizure or convulsion.

Sexual symptoms

Burning sensation in sexual organs or rectum (other than during intercourse), sexual indifference, impotence.

Female reproductive symptoms judged by the person to occur more frequently or severely than in most women. Painful menstruation, irregular menstrual periods. Excessive menstrual bleeding.

Differential Diagnosis

This disorder has to be differentiated from

(i) Physical disorders: Multisystem disorders e.g. SLE, Amyloidosis, Endocrinopathies etc.
(ii) Schizophrenia with multiple somatic delusions
(iii) Dysthymic disorder (Depressive neurosis) and Generalised anxiety disorder
(iv) Panic disorder
(v) Other somatoform disorders
(vi) Factitious disorder with physical symptoms.

Treatment

Non-invassive supportive long-term care of somatisation disorder patients that largely focussed on containing the use of medial resources and avoiding unnecessary surgery and medication use is required. Supportive techniques rather than intensive interpretive psychotherapy is recommended.

II. CONVERSION DISORDER

This syndrome is also characterised by the presence of multiple somatic complaints, but unlike other somatoform disorders, symptoms develop abruptly and cease to exist suddenly. It is also known as Hysteria. Most common presentation is Pseudoneurological symptoms e.g., convulsions, headache in response to stress. It is more common in females and in those societies where persons emotion are not allowed to be expressed. According to psychological theories, repressed emotions/wishes are converted into the bodily symptoms and thus the patient is able to seek attention. However, such process works outside patients awareness/consciousness. Therefore in such situations, he/she should not be blamed.

Treatment

Since, the patient is acting on the repressed emotions to seek attention, outside consciousness, aversive therapies e.g., ammonia sniffing, putting nasogastric tube should not be used. Such aversive therapies further damage the already fragmented emotions.

Rather, in such situations, patients should be provided with emotional support. He should be encouraged to ventilate his feelings. Importantly, once it becomes clear that symptoms belong to conversion reaction, and not due to any medical disorder, they should be overlooked.

Physician should take a non-confrontational approach and patient should not be given any explanation for his/her symptoms at the first meeting.

Environment of the family that led to such symptoms should be manipulated. Members should be encouraged to be open and discuss their emotions with each other.

III. HYPOCHONDRIASIS

The unrealistic fear or belief of having a serious disease persists despite medical reassurance and causes impairment in social or occupational functioning.

This disorder is commonly seen in general medical practice (ranging from 3 to 10%). Most commonly the age at onset is in adolescence, although frequently the disorder begins in the 30s for men and the 40s for women. The disorder is equally common in men and women.

Estimates indicate that 85% of all cases of hypochondriasis are of the *secondary* (depression being the commonest cause) and 15% of a *primary* nature. In its secondary forms, hypochondriasis represents a frequent feature of depression, anxiety, schizophrenia and early phases of dementia. It is also a common response after recuperation from any life-threatening illness (e.g. Coronary bypass etc.).

Headache, chest and gastrointestinal complaints and musculoskeletal symptoms are noted more frequently.

Complications are secondary to efforts to obtain medical care.

Differential Diagnosis

Hypochondriasis has to be differentiated from:

(i) True organic disease.

(ii) Psychotic disorder.

(iii) Others. Dysthmic Disorder, Generalised Anxiety Disorder, Panic Disorder, Obsessive Compulsive Disorder and Somatisation Disorder.

Management

It is necessary to exclude organic pathology. If it is secondary to primary illness e.g. depression, then by treating this, hypochondriacal symptoms will fade away. Treatment is limited to support and avoidance of continuous discussion of the patient's symptoms. The hospitalisations, tests and medications with addictive potential should be avoided.

IV. SOMATOFORM PAIN DISORDER

It is characterised by a clinical picture in which the predominant feature is the complaint of pain, in the absence of adequate physical findings and in association with evidence of the etiological role of psychological factors. The disturbance is not due to any other mental disorder.

Epidemiology

This disorder is common in general medical practice and is more frequently diagnosed in women. This disorder can occur at any stage of life but begins most frequently in adolescence or early adulthood.

Etiology

Severe psychosocial stress is a predisposing factor.

Clinical Picture

The 'doctor shopping,' excessive use of analgesics without relief for pain, requests for surgery and the assumption of an invalid role is common. The individual usually refuses to consider the role of psychological factors in the pain.

Table 1: Comparison of Physical (Organic) and Psychological Abdominal Pain

		Abdominal pain	
		Physical	*Psychological*
(i)	Frequency	10%	90%
(ii)	Pain	Localised	Diffuse
(iii)	Pain wakes at night	Often	Rarely
(iv)	Pain elsewhere in body	Unusual	Common
(v)	Vomiting	±	±
(vi)	Emotional state	Usually normal	Usually anxious, tense, sad
(vii)	Abnormalities on exam. and investigations	Positive	Negative
(viii)	Mannkopf's sign*	Positive	Negative

*(Pressure over or movement of painful part may lead to temporary increase in pulse rate by 10–30/min.)

Differential Diagnosis

Organic pain, somatisation disorder, depressive disorders, schizophrenia, malingering or pain associated with muscle contraction, headaches (Tension headaches).

Management

Complete disappearance of pain through suggestion, hypnosis or narcoanalysis suggests psychogenic pain disorder. The use of narcotic analgesics or chronic use of an anxiolytic drug should be avoided.

V. BODY DYSMORPHIC DISORDER

Patient is preoccupied with an imagined defect in his appearance. His concern increases if a slight physical defect is also present. These symptoms markedly affect patient personal, social and occupational functioning.

Treatment

SSRI are the mainstay of treatment. However, if it reaches delusional level, than antipsychotics may be given. Cognitive behaviour therapy also helps.

VI. MALINGERING

It is defined by ***intentional*** production of physical or psychological symptoms motivated by identifiable "external incentives" (avoiding work or military obligation, obtaining financial compensation, evading criminal prosecution, obtaining drugs etc). In contrast to factitious disorder, there should be an ***identifiable goal*** for behaviour other than that of securing the role or parenthood.

A high index of suspicion of malingering should be aroused if any combination of the following is noted:

(i) Medico-legal context of presentation e.g., the person's being referred by the court to the physician for examination.

(ii) Marked discrepancy between the person's claimed distress or disability and the objective findings.

(iii) Lack of cooperation with the diagnostic evaluation and prescribed treatment regimen.

(iv) The presence of antisocial personality disorder.

The detection of malingering is difficult, and each medical speciality has tended to develop its own set of guidelines for detection. Symptom relief in malingering is not often obtained by suggestions, hypnosis or intravenous barbiturates as it frequently is in conversion disorder.

10

Headache in General Practice: What you must know?

Headache is a common complaint that the primary care physician usually come across. Epidemiological studies suggest that headache is widely prevalent in the community and its lifetime prevalence varies from 5% to 83% in different samples depending upon the age, gender, place of the study and other criteria for the sample population. Sometimes headache may be a diagnosis in itself, while on the other time, this could just be an epitome indicating the proper assessment of the underlying illness. However, in headache clinics, primary headaches predominate and secondary headaches are relatively uncommon.

In this chapter we will discuss regarding common causes of headaches, their classification, approach to a patient with headache and lastly the treatment of individual headaches.

Pathogenesis of Headache

Headache is essentially a cerebral pathology and it originates when the sensory nerves in the cranial cavity are irritated. It must be kept in mind that only the dura and cranial vessels are pain sensitive. Brain parenchyma is pain insensitive and any damage to the brain itself is painless.

Hence, the pathologies that affect the cranial vessels and dura produce pain. Most commonly the pain occurs due to inflammatory pathologies like vasculitis and infections. Dura is sensitive to stretching and hence, pathologies that increase intracranial pressure suddenly e.g. cerebral haemorrhage, trauma etc. also induce pain. This is also the reason why most of the cerebral tumours are painless unless they are large enough to produce stretching.

Another issue to be remembered in these patients is the nerve supply of the scalp and cranium. Largely, anterior half of the head is supplied by sensory branches of trigeminal nerve while posterior half by upper cervical nerves. To make the issue more complicated, most of the facial structures—oral cavity, sinuses, ear, nose, eye, teeth etc. are supplied by sensory twigs from fifth cranial nerves. Hence, pain from the pathologies of these structures frequently refers to head.

Causes of Pain

This knowledge will help us in understanding the causes of headache. Headache can be classified into two broad categories—primary and secondary. Primary headache may be termed as functional and are similar to other psychiatric illnesses as available laboratory investigations—biochemical, pathological as well as neuro-radiological do not show any abnormality. These headaches originate in the brain due to activation of central pain perceiving structures. Common types of primary headaches are—migraine and its variants; tension type headache; headache with cranial autonomic symptoms e.g. cluster headache, hemicranias continua etc. Besides these three major primary headaches, there are a number of other primary headaches such as primary stabbing headache, medication overuse headache etc. and list is growing everyday when new headaches are being described.

In the second category, those headaches are included which arise due to some other pathology inside the body. Usually these pathologies affect the facial or the cranial structures. Common causes of secondary headaches are described in Table 1.

Table 1: Causes of Secondary Headaches

S.N.	*Causes*	*S.N.*	*Causes*
1.	*Cerebral causes*	3.	*ENT causes*
	Trauma to the head		ASOM
	Carebrovascular accidents		Mastoiditis
	Intracranial haemorrhage		Acute rhinitis
	Cerebral		Acute sinusitis
	Subdural		Pharyngitis
	Epidural		Contact point headache
	Sub-arachnoid	4.	*Ophthalmic causes*
	Cerebral		Conjunctivitis
	Intracranial infections		Iritis
	Meningitis		Iridocyclitis
	Encephalitis		Glaucoma
	Cerebral abscess vasculitis		Refractory errors
	Giant cell arteritis		Intra-orbital mass

Contd...

S.N.	Causes	S.N.	Causes
	Hydrocephalus	5.	*Dental causes*
	Idiopathic intracranial hypertension		Dental infections
	Cranial tumours		Periodontal abscess
2.	*Cervical causes*	6.	*Other general causes*
	Spondylitis		Malignant hypertension
	Compression fracture		Hypothyroidism
	PIVD		Fever
	Nerve entrapment		

Approach Towards the Patient

Even when we know that most of the time headaches are primary, it is very important to rule out the secondary headaches, as misdiagnosis may result in spread of primary pathology.

Like any other illness, it is prudent the patient should be encouraged to provide as much details as he can. If the patient is unable to provide details adequately, clinician may guide the interview to find out the relevant information. But care should be taken not to ask leading questions early in the interview for two reasons—this will prejudice the clinician regarding a diagnosis and more importantly, patient often have co-morbid illnesses that may be masked.

You must try to collect details regarding total duration since the patient is suffering from headache, average duration of each headache episode, character, frequency and intensity of headaches, any change in the character of headache, location of headache, radiation and referred pain, precipitating, aggravating, relieving factors and lastly the associated symptoms.

Long history of headache that are stereotyped usually points towards presence of primary headaches, whereas recent onset headache or a change in its character may be a guide towards secondary headache. Similarly, duration of headache episode is very important as tension type headache may last from 30 minutes to seven days while usual duration of migraine is from four hours to four days. Cluster headache usually lasts for hours only. Secondary headaches are of variable duration and that depends upon the underlying pathology. If the pathology is severe, headache may persists until therapy is given, while headache produced by third ventricular tumours, lasts for the time till they block the interventricular foramen and may be relieved by change in position. Similarly, character of headache is very important as tension type headache and most of the other secondary headache may have band-like or heaviness like pain. Cluster headache is often boring type and the same character is borne by headache due to meningeal irritation e.g., sub-arachnoid haemorrhage or the meningitis and glaucoma as well. Classically migraine is pulsating or throbbing type and it was described to increase with each heart beat.

Frequency may give you a clue regarding progression of disease and help in initiation of proper management. Frequent primary headaches may warrant a prophylactic therapy while infrequent but severe headaches may require just the episodic treatment. Intensity varies among the headaches and headache due to acute SAH is thought to be most severe while tension type headache is mild and rarely moderate. Migraine as well as cluster headache are moderate to severe. Pain due to glaucoma, iridocyclitis are severe, while that of conjunctivitis, rhinitis or sinusitis is mild to mild. Headache due to increased intracranial tension is usually mild.

Although primary headaches often change their character with time, even than this must alert the clinician to seek the presence of another primary or secondary headache.

Location of headache often provides important clues towards the diagnosis. Migraine is most commonly temporal or retro/intra-orbital and the cluster headache is most prominently orbital. Migraine may present with pain in the neck, just below the occiput and this may be unilateral.

Pain in secondary headaches is most severe at the local area though it may refer to different areas depending upon the nerve supply. In general, pain from supratentorail compartment and facial structure is referred to trigeminal territory while pain from the posterior fossa is referred to area supplied by upper cervical roots.

Associated symptoms are most important clue to the diagnosis. Tension type headache usually do not have any associated symptom while migraine often presents with nausea-vomiting, phonophobia or photophobia. Cluster headache patients show cranial autonomic symptoms e.g., ipsilateral miosis, eyelid swelling, lacrimation, conjunctival injection, nasal stuffiness or watering. Patient should be encouraged to describe these symptoms in detail as they are most important to the diagnosis of secondary headaches. Fever, delirium, focal neurological deficits, convulsions are seen in cerebral pathologies; change in hearing, vertigo, ear discharge, dysphagia, nasal blockade, fever may point to otolaryngological illnesses; haloes, red eye, change in vision, pain on movement of eye may be a clue to eye diseases and gum swelling and problem in bite are manifestations of dental problems. The list is not exhaustive and reader is advised to read the specialised literature for better access to the knowledge.

Investigations

In general, if the physician is sure that headache is benign and is primary headache, then neuroradiological examination is not required. However, some people prefer a safer approach and they order neuroradiological examination. According to the expert's opinion this is mandatory in following conditions:

1. First onset of headache
2. Recent change in character of headache
3. When patient presents with some neurological deficits
4. Sudden severe intolerable headache
5. Patient's preference/assurance scan

When secondary headaches are suspected, relevant laboratory tests must be ordered. It must be noted here that some patients may not have any obvious clinical sign despite gross cerebral pathology.

Treatment

It is not possible to provide the detailed account of the treatment here. So we are providing general guidelines. More interested readers may refer to the specialised textbooks of headache.

Treatment depends upon the type of headache. For secondary headaches, appropriate therapy must be instituted at the earliest to prevent the spread of illness. Before starting the therapy, diagnosis must be confirmed by appropriate investigations.

For occasional primary headaches, simple analgesics are preferred. But for acute severe attack of migraine and cluster headache, ergots and triptanes can also be used taking appropriate precautions as both these agents are vasopressors and vasoconstrictors. It must be remembered that patient frequently misuse these drugs and sometimes, doctors also prescribe these drugs for chronic use which can result in development of medication-overuse headache, which is often refractory to the treatment. In general, analgesics, ergots and triptanes must not be used more than 4–5 days a month. If a patient is complaining of more frequent headache, prophylactic drugs must be started. Acute attack of cluster headache most often responds to the hyperbaric oxygen therapy.

For prophylaxis of headache a number of drugs are available e.g., antidepressants, valproate, lamotrigine, zonisamide, topiramate, cyproheptadine, corticosteroids, beta-blockers etc. Choice of drug depends upon the physician's preference, frequency, severity of headache, adverse effects and co-morbid conditions.

Here I wish to emphasise the fact that primary headache patients frequently have co-morbid psychiatric illnesses e.g., mood disorders, anxiety disorders, sleep disorders and personality disorders. Unless these disorders are adequately addressed, remission is impossible. Hence, all primary headache patients must be provided due consideration to manage the psychiatric illness both pharmacologically as well as through non-pharmacological means. Since, primary headaches are chronic disorders, avoidance of precipitating and aggravating factors is of paramount importance. Its importance must not be underestimated.

11

Problems of Sleep

Many of us at different times in our life have experienced that we are not able to sleep properly. Proper sleep is considered one that is adequate in duration, is good quality and refreshing. Among all these factors, refreshing sleep is one that needs emphasis as duration is individualised and cannot be generalised beyond a certain limit. Quality of sleep also varies and is dependent upon the corporal and environmental factors. Since quality of sleep differs from person to person and is dependent upon one's expectations regarding sleep, hence 'refreshing' sleep is the surrogate marker. Refreshing means that one is able to work in best of his capacity after leaving the bed provided he is not having any other problem. In essence, one must seek the advice from the physician when he is not able to get the refreshing sleep.

Why is he not having refreshing sleep?

This is the key question that every sleep-specialist looks for whenever you approach him for these complaints. Though the issue appears trivial, but most of the times it is very difficult to answer owing to multiple reasons. First, most of us want the readymade therapy (sleeping pills) and usually do not help the physician to go into details. Second, we are ignorant regarding the development of complaints and hence are not able to provide the exact and adequate information; probably getting a good quality sleep is very low in our priority list despite the fact that non-refreshing sleep may lead to other pathologies e.g. cognitive and memory problems, work performance, mood regulation, stress, headache, hypertension, diabetes etc. Third, a number of people (including physicians) are not aware that non-refreshing sleep is sometimes the only complaint of other serious underlying disorders e.g. depression, obstructive sleep apnea, periodic limb movement disorder, cardiac decompensation and other respiratory disorders. Fourth, we are not courageous enough to accept the fact that we are doing something wrong with our body and hence try to hide those facts e.g. not following sleep hygiene, consumption of additive substances etc. Besides these there are other factors why sleep complaints get unrecognised.

How common is this problem?

Though at present we do not have any data regarding sleep problems in Indian adult population, data regarding adolescent population of a metropolitan city is available which shows that majority of adolescents are sleeping one hour less than required duration in the respective age and it is culminating in their school performance. Western data show that at a given time around 15% persons in a given community suffer from insomnia.

What causes this problem?

There are a number of reasons that contribute to the development of non-refreshing sleep. The causative factors are not mutually exclusive and most of the times they are present together to compound the diagnostic and therapeutic exercises. Environmental factors e.g. extremes of temperature, poor sleeping place, overcrowded room, mosquitoes, noisy surroundings, poor ventilation, bright lighting etc. may cause non-refreshing sleep. Medical disorders e.g. pain anywhere in the body, difficulty breathing in lying down situation, certain brain tumours etc.; psychiatric disorders e.g. primary insomnia, depression, stress, anxiety etc; frequently changing work shifts or time zones, strange surroundings etc.; substance dependence e.g. sleeping pills, alcohol, cannabis, cocaine etc. may cause or contribute to the illness. Hence, remember it, pills are not the answer always and you need to help your treating physician to reach a conclusion and address underlying issues.

From medical point of view, we divide these factors into three classes—predisposing, precipitating and perpetuating. Predisposing factors are those that you are born with and most of the times you may not have control over them e.g. your body clock which needs less sleep, your arousal level e.g. your heart beats faster, higher metabolic rates, you are easily disturbed by the trivial emotional stimuli, and you keeps on thinking a lot on trivial issues. However, recent research has contradicted this issue and found these factors to be the result rather than cause of insomnia. Researchers have noticed that insomnia patients keep worrying about their sleep-problem and its effect on their functioning, hence further increasing the stress and consequent insomnia. Secondly, these patients keep on lying on their bed awake in the night and slowly, their bed and bedroom becomes associated with awakening than sleeping (conditioning). A good therapist tries to curtail the impact of modifiable factors but it cannot be done without cooperation of the patient and sometimes co-sleeper.

Precipitating factors are those that are under our control and which when present, increase the chances of poor sleep in an any person, however, their impact is more on persons with predisposing factors. All environmental, situational, medical or psychiatric illnesses, medications etc. as mentioned above are included in this group. These situations usually induce episodic problems but in presence of perpetuating factors, lead to chronic problem.

Perpetuating factors include cognitive and behavioural factors. As we have mentioned, negative conditioning with the sleeping environment is an important issue. Insomnia sufferers in absence of proper guidance and quick relief adopt some antagonistic behaviours e.g. changing sleep schedule, sleeping pills, heavy intake of caffeine during waking period, performing activities in the bed for which it is not meant e.g. reading books, watching television, lying wake up in the bed waiting for the sleep etc.

Is this problem serious?

Seriousness of any problem cannot always be judged by the overtly conspicuous life-threatening issues or physical debility produced. When we see it at large scale, loss of working hours, poor work capacity, and increased health related expenditure and in turn, resultant loss of money gives better idea regarding gravity of any issue. It is estimated that in US around 2 billion dollars are spent every year on sleeping pills and indirect cost from above mentioned reasons when added, the figure reaches more than 100 billion US$ per annum. Though we do not have Indian data at present, but situation is not better than that mentioned above, at least in metropolitan cities. In addition, we wish to again emphasise upon the cost of insufficient sleep borne by you in terms of cognitive and memory problems, work performance, mood regulation, stress, headache, hypertension, diabetes etc. and their management.

How to combat it?

Fighting with insomnia is the joint work of physician and patient. Both must be willing and dedicated to diagnose the underlying problem and treat it. Sometimes, what your physician suggests may not appeal to you, but you must talk to him more about your queries and fears at that time only to help yourself. Always remember, pills are usually not the answer to chronic insomnia and minute change in environmental, cognitive and behavioural factors may be more useful.

The evaluation starts with the information regarding your complaints and includes complete physical examination, systemic examination, psychiatric evaluation and sometimes laboratory investigations. We will not go into details of all these.

Therapy of insomnia includes management of medical disorders, psychiatric illness and sometimes sleeping pills. Multiple therapeutic options are available and choice of any of these depends upon the physician's preference and his judgement. We will not discuss these in detail as they are out of scope for this book but will discuss sleep promoting behaviours.

Sleep promoting behaviours are also called, good sleep hygiene and they are helpful not only therapy of the insomnia but also its prevention. These are mentioned in table 1.

Table 1:

Do's	*Don'ts*
• Maintain a regular sleep-wake schedule	• Do not take caffeinated beverages/coffee/tea before four hours of bed
• Should take adequate amount of food before going to bed	• Do not indulge in heavy exercise before going to bed
• Indulge in physical exercises during the day	• Avoid daytime naps
• Take balanced diet	• Surrounding environment should not be crowded, very cold or hot, noisy etc.
• Go to bed only when sleepy	• Avoid alcohol close to bedtime
• Always leave the bed at the same time in the morning	• Avoid doing your work in the bed
	• Should not be hungry or do not eat too much in the dinner
	• Do not watch TV while in bed

Stimulus control therapy

As we have discussed above, many a times bed becomes negatively associated with the sleep and thus person remains wake up while in bed. Stimulus control therapy aims at deconditioning of negative cues and reconditioning of healthy cues. However, this requires dedication from the patient side, and shows its effect in at least 1–2 weeks, during which the anxiety and sleep complaints may actually increase. Hence, you must be ready to pay initial price for long-term benefits. This therapy consists of simple instructions, but picking up the right patient is probably the most important part of therapy.

In this therapy, patient is ask to follow simple instructions as follows: (a) Use the bed for sleeping only; (b) Leave the bed when not able to sleep; (c) Indulge in relaxing exercises while out of bed and lastly, when feels sleepy, go to bed. This therapy has proven very effective when instituted in the right candidates and does not include any kind of pill-popping.

Sleep-restriction therapy

This therapy increases the internal sleep pressure and thus helps in achieving the normal sleep wake cycle. In this therapy, patient's bed time is reduced to the amount of time spent in sleep, which is decided by the sleep-diary. For example, if you spend 8 hours in bed and sleep for five hours only, your time in bed is reduced to 5 hours only. When you are able to sleep for approximately 90% of your time in bed (determined by sleep diaries), time in bed is increased gradually. This is very effective method, though in the initial part, it may rather provoke the anxiety and may interfere with sleep.

Cognitive behaviour therapy

This therapy addresses the unrealistic beliefs and attitudes regarding sleep and sleep problems. In coordination with the patient, these beliefs are identified and then challenged gradually. This therapy can be combined with any of the above mentioned methods to achieve total control over the situation.

Pharmacological management

There are many drugs available that can induce sleep, to name a few: benzodiazepines, non-benzodiazepine hypnotics, some antidepressants, centrally acting antihistaminics etc. Whether a drug should be given or not, and if yes, which is the right drug, depends upon the condition of the patient and diagnosis. Since, many of these drugs are available over the counter and sometimes, unfortunately these are prescribed indiscriminately by some medical persons, they are overused and sometimes in the wrong patient, leading to mismanagement and long-term deterioration of underlying problem. We wish to emphasise here that the benzodispines at least, are not effective pharmacologically after 2–3 weeks of continuous use. Hence, there is no point in continuing them beyond that period.

12

Stress and its Management

Stress is what everybody experiences in daily life. Stress refers both to a subject, and a predicate, an event and the consequences of an event. Stress was coined by Hens Seyle. For the sake of simple understanding , we take the example of stress as a false alarm in the brain which is activated in response to a danger signal. Stress is actually a good thing to have. The biological purpose of stress is to prepare us fight real, physical danger. When the danger alarm is turned on, it produces a physiological response called the "fight or flight" reaction, which helps us to fight the danger or flee it.

Stress at work is a relatively new phenomenon of modern lifestyles. It covers all professions, starting from an artist to a surgeon, or a commercial pilot to a sales executives, workers, consequently, affects the health of the individuals.

Any change in the environment demands some coping; and little stress helps us adapt. But, beyond some point stress becomes distress basically individual's response to fight/flight a situation. Stress when pathological, causes lot of physical and mental health problems. There are two main types of stress experienced by humans, either chronic or that which is emergency-induced. The chronic type of stress can be particularly harmful to the brain because of hormones and chemicals referred to as glucocorticoids or GCs. When the body experiences a rush of adrenaline which is accompanied by stress, a portion of our brain called the adrenal cortex begins to release these GCs which are useful for dealing with the emergency-type of stressors.

Etiology of stress

The stress activate the pituitary adrenal axis and lead to what is known as the General Adaptation Syndrome (given by Hens Seyle in 1945). It consists of:

(i) Alarm reaction (shock), which causes the body defenses to become alert about the danger or stress.

(ii) Resistance (adaptation to stress) which is characterised by secretion of hormones or defensive chemicals to contain stress.

(iii) Exhaustion (resistance to prolonged stress cannot be maintained), which results in failure of body defenses when stress in severe and prolonged.

George Engel postulated that in the stressed state, all neuroregulatory mechanisms undergo functional changes that depress the body's homeostatic mechanisms, leaving the body vulnerable to infection and other disorders.

Neurophysiological pathways that mediate stress reactions are cerebral cortex, limbic system, hypothalamus, adrenal medulla, sympathetic and parasympathetic nervous system. Neuromessengers include such hormones cortisol, thyroxine, epinephrine. Small amounts of stress enhance immune function, excess stress impairs it.

Stress and psychiatric disorders

The expression of mental illnesses was affected by life circumstances. Meyer proposed the biopsychosocial model. Individual's response to stress is modified by a number of intrinsic factors (e.g. genetic vulnerability, premorbid personality) and extrinsic factors (e.g. social support). The model incorporated individual temperamental and experiential characteristics such as potential vulnerability (or resiliency) factors, stressful life events as initiating or exacerbating factors, and a variety of support networks as modifying factors for the occurrence of mental illness.

Stress plays a role in a number of psychiatric disorders.

1. Major depressive disorder – Several life events predict onset of major depressive disorder.
2. Chronic interpersonal stress is an important risk factor for relapse in Schizophrenia. Several studies had shown that high degree of expressed emotions cause more relapse than those who live in families with low expressed emotions.
3. Anxiety disorders – Panic disorders has its onset with stressful life events. Interpersonal conflict or serious illness has been shown to trigger the onset of panic disorder in susceptible individual.
4. Post traumatic stress disorder and acute stress reaction had their onset following traumatic events such as violent assaults or serious accidents.

Stress and non-psychiatric disorders

Stress plays an important role as a triggering factor or as an exacerbating factor in a number of non-psychiatric disorders like angina, arrhythmias, coronary spasms, asthma, connective tissue diseases like systemic lupus erythematosus, rheumatoid arthritis, headaches, hypertension, hyperventilation syndrome, inflammatory bowel diseases like Crohn's disease, irritable bowel syndrome, ulcerative colitis, metabolic and endocrine disorders, neurodermatitis, obesity, osteoarthritis, peptic ulcer disease, Raynaud's disease, syncope, uriticaria, angioedema.

Stress and medical professionals

1. Do doctors experience stress? If so, what could be the contributing factors?

   Long working hours, increased expectations of clients, dealing with emotions—constant struggle of life and death, lack of communication skills, etc.

2. Is medical profession more stressful?

   Low salary, increased work pressures, increased duration of training, poor quality of training, stress relating to the work-home interference increased during their early career. Neurotic traits, conscientiousness, and lack of support from one's partner and colleagues, appeared to be predictive of stress.

   Some studies found that subjective ratings of high pressure and insufficient sleep are associated with poor job performance in medical residents.

   Unique challenges facing female residents include the existence of gender bias and sexual harassment, a scarcity of female mentors in leadership positions, and work/family conflicts.

3. What must they do as good doctor?

   Self defeating beliefs – Very often, doctors blame themselves for difficulties. Long working hours, high degree of professional commitment, no vacations, self defeating beliefs, poor social support often lead to burn out. Maladaptive coping style in which the individual perceive their environment as stressful.

   Burn out syndrome is an illness seen among the high achieving intellectual workers like teachers, doctors, and IT consultants appeared to be victims. The clinical criteria includes physical, mental, and emotional exhaustion, uneasiness, and lack of empathy. Burn out was classed as an illness affecting those with low self reliance and depressive personality types, primarily of the female gender.

STRESS MANAGEMENT

There are five basic skills that form core of all stress management programmes: self observation, cognitive restructuring, relaxation training, time management, and problem solving.

Self Observation

Stress Diaries are important for understanding the causes of short-term stress in your life. They also give you an important insight into how you react to stress, and help you to identify the level of stress at which you prefer to operate.

The idea behind Stress Diaries is that, on a regular basis, you record information about the stresses you are experiencing, so that you can analyse these stresses and then manage them.

Stress Diaries help you to understand:

- The causes of stress in more detail;
- The levels of stress at which you operate most effectively; and
- How you react to stress, and whether your reactions are appropriate and useful.
- They establish a pattern that you can analyse to extract the information that you need.
- Also make an entry in your diary after each incident that is stressful enough for you to feel that it is significant.

Every time you make an entry, record the following information:

- The date and time of the entry.
- The most recent stressful event you have experienced.
- How happy you feel now, using a subjective assessment on a scale of –10 (the most unhappy you have ever been) to +10 (the happiest you have been). As well as this, write down the mood you are feeling.
- How effectively you are working now (a subjective assessment, on a scale of 0 to 10. A 0 here would show complete ineffectiveness, while a 10 would show the greatest effectiveness you have ever achieved.
- The fundamental cause of the stress (being as honest and objective as possible).

You may also want to note:

- How stressed you feel now, again on a subjective scale of 0 to 10. As before, 0 here would be the most relaxed you have ever been, while 10 would show the greatest stress you have ever experienced.
- The symptom you felt (e.g. "butterflies in your stomach", anger, headache, raised pulse rate, sweaty palms, etc.).
- How well you handled the event: Did your reaction help solve the problem, or did it inflame it?

Analyse the diary at the end of this period.

Analysing the Diary

Analyse the diary in the following ways:

- First, look at the different stresses you experienced during the time you kept your diary. List the types of stress that you experienced by frequency, with the most frequent stresses at the top of the list.
- Next, prepare a second list with the most unpleasant stresses at the top of the list and the least unpleasant at the bottom.
- Looking at your lists of stresses, those at the top of each list are the most important for you to learn to control.
- Working through the stresses, look at your assessments of their underlying causes, and your appraisal of how well you handled the stressful event. Do these show you areas where you handled stress poorly, and could improve your stress management skills? If so, list these.
- Next, look through your diary at the situations that cause you stress. List these.
- Finally, look at how you felt when you were under stress. Look at how it affected your happiness and your effectiveness, understand how you behaved, and think about how you felt.
- Having analysed your diary, you should fully understand what the most important and frequent sources of stress are in your life. You should appreciate the levels of stress

at which you are happiest. You should also know the sort of situations that cause you stress so that you can prepare for them and manage them well.

- As well as this, you should now understand how you react to stress, and the symptoms that you show when you are stressed. When you experience these symptoms in the future, this should be a trigger for you to use appropriate stress management techniques.

Cognitive restructuring

Cognition plays a key role in the stress and coping process. Cognitive behavioural therapy makes the individuals aware of and to change their maladaptive thoughts, beliefs, and expectations.

Relaxation training

Relaxation skills are very helpful in managing stress. When individuals learn to relax, their overall muscle tension is reduced, as is their overall level of automatic arousal. Individuals who are able to relax are also more likely to be able to think more rationally and be able to restructure negative cognitions when faced with stressful events. Finally, relaxation skills like Jacobson are found to be quite effective.

Other alternative measures

Yoga, acupuncture, acupressure, laughter, breathing exercises, aerobic exercises, music therapy, massaging etc. and several techniques have been found to be of considerable use in stress management.

Adopting a humorous view towards life's situations can take the edge off everyday stressors. Not being too serious or in a constant alert mode helps to maintain the equanimity of mind and promote clear thinking. Being able to laugh stress away is the smartest way to ward off its effects.

A sense of humour also allows us to perceive and appreciate the incongruities of life and provides moments of delight. The emotions we experience directly affect our immune system. The positive emotions can create neurochemical changes that buffer the immunosuppressive effects of stress.

TIME MANAGEMENT

The old saying goes- "Time is money." We must learn to focus ourselves on most urgent and most important areas. The alternative is to work more intelligently, by focusing on the things that are important for job success and reducing the time we spend on low priority tasks. Job analysis is the first step in doing this.

By understanding the priorities in your job, and what constitutes success within it, you can focus on these activities and minimise work on other tasks as much as possible. This helps you get the greatest return from the work you do, and keep your workload under control.

Problem Solving

Problem solving involves few basic steps

1. Problem identification
2. Generating alternatives
3. Evaluating alternatives and finding the best solution.

Some simple practices can help minimise stress

- Sit straight and comfortably on your seat, and try breathing exercises. It will relax your nerves and muscles.
- Relax and count backwards (20, 19, 18, 17, 16, 15....)

Are You in Danger of Burning Out?

If you feel that you are in danger of burning out, the suggestions below can help you correct the situation:

- Re-evaluate your goals and prioritise them.
- Evaluate the demands placed on you and see how they fit in with your goals.
- Identify your ability to comfortably meet these demands.
- If people demand too much emotional energy, become more unapproachable and less sympathetic. Involve other people in a supportive role. Acknowledge your own humanity: remember that you have a right to pleasure and a right to relaxation.
- Learn stress management skills.
- Identify stressors in your life, such as work, or family. Get the support of your friends, family and even counselling in reducing stress.
- Ensure that you are following a healthy lifestyle:
  1. Get adequate sleep and rest to maintain your energy levels.
  2. Ensure that you are eating a healthy, balanced diet—bad diet can make you ill or feel bad. Limit your caffeine and alcohol intake.
  3. Try to recognise your spiritual needs that may have been buried under the mires of worldly pursuits.
- Develop alternative activities such as a relaxing hobby to take your mind off problems.

13

Psychosexual Disorders

Individuals with sexual disorders are likely to be encountered by clinicians in all disciplines. They may present to venereologists, surgeons, psychiatrists, physicians, gynecologists or so-called sexologists. It is important for all clinicians to take a detailed sexual history from their patients and have an understanding of the diagnosis and management of the sexual disorders.

The name for this diagnostic class emphasises that psychological factors are assumed to be of major etiologic significance in the development of the disorder. Disorders of sexual functioning that are caused exclusively by organic factors, even though they have psychological consequences, are not discussed in this chapter. The topics which will be discussed in this chapter are as follows:

- Normal human sexual response cycle (Table 1)
- Various disorders according to individual phase (Table 2) and their management e.g., Erectile Dysfunction (Table 3) and Dhat Syndrome
- Paraphilias
- Other sexual disorders: Gender identity disorder and homosexuality

Normal human sexual response cycle can be divided into various phases, which are found in almost all people in the sequential manner. Table 1 summarises various phases.

The sexual disorders can be classified according to stage in which they occur. This differentiation is important in understanding of pathology as well as the therapy. While collecting the information/history it is important to rule out dysfunction in every stage, sequentially starting from stage 1, since a disorder in the earlier stage can precipitate/aggravate or mimic problem of subsequent stage. For example, patient may present with complaints of erectile dysfunction while he is actually suffering from hypoactive sexual desire disorder. In such case investigations or treatment of erectile disorder will result in unnecessary investigations or therapeutic trials.

Various sexual disorder are classified according to stages of response cycle in Table 2.

Table 1: Normal Human Sexual Response Cycle

A normal human sexual response cycle can be divided into five phases.

Males	*Females*
1. Appetitive Phase This is the phase which occurs before the actual sexual response cycle. This consists of sexual fantasies and a desire to have sexual activity.	**1. Appetitive Phase** This is the phase which occurs before the actual sexual response cycle. This consists of sexual fantasies and a desire to have sexual activity.
2. Excitement Phase This is the first true phase of the sexual response cycle, which starts with physical stimulation and/or by appetitive phase. The duration of this phase is highly variable and may last for several minutes (or longer). The major changes during this phase are listed below: • Penile erection, due to vascocongestion of corpus cavernoa. • Elevation of testes with scrotal sac.	**2. Excitement Phase** This is the first true phase of the sexual response cycle, which starts with physical stimulation and/or by appetitive phase. The duration of this phase is highly variable and may last for several minutes (or longer). The major changes during this phase are listed below: • Lubrication of vagina by a transudate. • Erection of nipples (in most women). • Erection of clitoris. • Thickening of labia minora.
3. Plateau Phase This is an intermediate phase just before actual orgasm, at the height of excitement. It is often difficult to differentiate the plateau phase from the excitement phase. The duration of this phase may last from *half to several* minutes. The following important changes occur during this phase: • Sexual flush (inconsistent). • Autonomic hyperactivity. • Erection and engorgement of penis to full size. • Elevation and enlargement of testes. • Dew drops on gland penis (2–3 drops of mucoid fluid with spermatozoa).	**3. Plateau Phase** This is an intermediate phase just before actual orgasm, at the height of excitement. It is often difficult to differentiate the plateau phase from the excitement phase. The duration of this phase may last from *half to several* minutes. The following important changes occur during this phase: • Sexual flush (inconsistent). • Autonomic hyperactivity. • Retraction of clitoris behind the pupuce. • Development of *orgasmic platform* in the lower 1/3rd of vagina, with lengthening and ballooning of vagina. • Enlargement of breasts and labia minora, increased vaginal transudate.

Contd...

Males	*Females*
4. Orgasmic Phase This is the phase with a peak of sexual excitement followed by a release of sexual tension, and rhythmic contractions of pelvic reproductive organs. The duration of this phase may last from 3–15 *seconds*. The important changes are as follows: 14–10 contractions of penile urethra, prostate, vas, and seminal vesicles; at about 0.8 sec. intervals. • Autonomic excitement becomes marked in this phase. Doubling of pulse rate and respiratory rate, and 10–40 mm. Increase in systolic and diastolic BP occur. • *Ejaculatory inevitability* precedes orgasm. • Ejaculatory spurt (30–60 cm; decreases with age). • Contractions of external and internal sphincters.	**4. Orgasmic Phase** This is the phase with a peak of sexual excitement followed by a release of sexual tension, and rhythmic contractions of pelvic reproductive organs. The duration of this phase may last from 3–15 *seconds*. The important changes are as follows: • 3–15 contractions of lower 1/3rd of vagina, cervix and uterus; at about 0.8 sec. intervals. No contractions occur in clitoris. • Autonomic excitement becomes marked in this phase. Doubling of pulse rate and respiratory rate, and 10–40 mm. increase in systolic and diastolic BP occur. • Contractions of external and internal sphincters. The duration of this phase may last from 3–15 *seconds*.
5. Resolution Phase This phase is characterised by the following common features in both sexes: • A general sense of relaxation and well-being, after the slight clouding of consciousness during the orgasmic phase. • Disappearance of sexual flush followed by fine perspiration. • Gradual decrease in vasocongestion from sexual organs and rest of the body. • Refractory period for further orgasm in males varies from few minutes to many hours.	**5. Resolution Phase** This phase is characterised by the following common features in both sexes: • A general sense of relaxation and well-being, after the slight clouding of consciousness during the orgasmic phase. • Disappearance of sexual flush followed by fine perspiration. • Gradual decrease in vasocongestion from sexual organs and rest of the body. • While there is usually no refractory period in females.

Table 2: The Various Psychosexual Dysfunctions can be classified as

A.	**Sexual Response Phase**	**Related Dysfunctions**
I.	Appetitive	* Hypoactive Sexual Desire Disorder
		* Sexual Aversion Disorder
II.	Excitement	* Female Sexual Arousal Disorder
		* Male Erectile Disorder
III.	Orgasm	* Female Orgasmic Disorder (Anorgasmia)
		* Male Orgasmic Disorder (Retarded Ejaculation)
		* Premature Ejaculation
IV.	Resolution	* No Dysfunctions Reported
B.	**Others not Related to Sexual Phase**	**Related Dysfunctions**
I.	Sexual Pain Disorders	* Vaginismus (female)
		* Dyspareunia (male and female)
II.	Others	* Orgasmic anhedonia, Genital pain during masturbation, female analogue of premature ejaculation.

Dhat Syndrome

It is a true culture bound sex neurosis quite common in the natives of the Indian subcontinent (e.g. those living in India, Bangladesh, Sri Lanka, Myanmar, Nepal, Pakistan, etc.). Dhat syndrome can be properly conceptualised in its entirely with an adequate understanding of the word 'Dhat' and the certain ancient Indian beliefs about it. Similar syndromes exist in China (Shenkui) and Taiwan (Shen-k'uei).

History. The Sanskrit word '*Dhatu*', according to Susruta Samhita (an ancient treatise of Indian medicine), means the elixir which constitutes the body and has given rise to the term '*Dhat*'. According to Susruta Samhita as well as Ayurveda (the Indian system of medicine), the disturbances in the '*Dhatus*' have been elucidated in the Charak Samhita (another ancient treatise of Indian medicine). It describes a disorder resembling *Dhat* syndrome by the name '*Shukrameha*' in which there is passage of semen in the urine. According to Shiva Samhita (an epic in Hindu Mythology) the falling of seed (sperm) leads towards death, the keeping of one's seed is life. Hence with all his power should a man hold his seed. Though purists still use the term *'Dhatu'* to connote *'Paddarth'* or basic material or metal, the wider and colloquial use today is synonymous with semen. In India, semen is also known by the name '*Virya*' derived from a Sanskrit word meaning bravery, valour, strength, power or that which generates power and greatness. '*Shukra*' is a term which more specifically refers to the sperm content of seminal fluid and is derived from the Sanskrit word '*shuch*' which literally means the 'essence' or fire or glow.

Thus, the belief that the loss of semen from the body in any manner—masturbation, spermatorrhoea, nocturnal emission, premarital or extramarital relations, is the most harmful thing that can possibly happen and its preservation guarantees health and longevity. Sometimes an overvalued idea that the semen has become foul smelling and less viscous in consistency may be present.

Commonly Encountered Disorders

(A) Male Erectile Dysfunction is one of the commonest causes for seeking help. Various causes can be classified as follows (Table 3).

Table 3: Classification and Common Causes of Erectile Dysfunction

Category of erectile dysfunction	*Common causes*	*Pathophysiology*
Psychogenic	Performance anxiety, relationship problems, Psychological stress, depression etc.	Loss of libido, over inhibition, or impaired nitric oxide release
Neurogenic	Stroke or Alzheimer's disease Spinal cord injury Radical pelvic surgery Diabetic neuropathy Pelvic injury	Failure to initiate nerve Impulse or interrupted neural transmission
Hormonal	Hypogonadism Hyperprolactinemia	Loss of libido and inadequate nitric oxide release
Vasculogenic (arterial or cavernosal)	Atheroscerlosis hypertension Diabetes mellitus Trauma Peyronie's disease	Inadequate arterial flow or impaired veno-occlusion
Drug induced	Antihypertensive and antidepressant drugs Antiandrogens alcohol abuse Cigarette smoking	Central suppression, Decreased libido, Alcoholic neuropathy, Vascular insufficiency
Caused by other systemic diseases and ageing	Old age Diabetes mellitus Chronic renal failure Coronary heart disease	Usually multifactorial, resulting in neural and vascular dysfunction

The belief of so-called precious and life preserving property of semen is deeply ingrained in the Indian culture. The quacks and lay experts reinforce this traditional belief and thrive on the ignorance of the common man. Sometimes, this belief is perpetuated by the friends or relatives, who had suffered from this syndrome. 'Dhat Syndrome' a term first used by *N.N. Wig* in 1960, is a commonly recognised clinical entity in Indian culture.

Following points differentiate between Psychological and organic dysfunction.

Table 4: Differentiation between Psychogenic and Organic Erectile Dysfunction

	Feature	*Psychogenic*	*Organic*
I.	Age	Younger	Older
II.	Onset	Acute	Gradual (except trauma/surgery)
III.	Circumstances	Situational	Global
IV.	Precipitants	Psychogenic condition	Age, Vascular insufficiency, DM etc.
V.	Symptom course	Intermittent	Consistent or Progressive
VI.	Desire	Decreased	Normal
VII.	Organic risks	Absent, variable	Present
VIII.	Partner problem	At onset	Secondary
IX.	Anxiety and Fear	Primary	Secondary
X.	Erectile function		
	— before intromission	May be present	Usually absent except in pelvic steal phenomenon
	— before intromission	Variable with different partners	Usually absent
XI.	Erectile response to other sexual stimuli	Usually present	Usually absent
XII.	Nocturnal or morning erections	Initially present and full; lost in longstanding dysfunction	Absent or reduced in frequency and intensity
XIII.	Associated ejaculatory disorder	PME or intermittent loss of ejaculation	Retrograde or absent ejaculation
XIV.	Nocturnal Penile Tumescence (NPT)		
	• Total time (min/night)	>90–180	<60
	• Circumferential change	0.2 cm	2 cm
	• Penile brachial index (PBI)	>0.70	<0.60
	• Bulbocavernus reflex latency	< 35 sec	> 40 sec

Following table shows treatment options for male erectile dysfunction.

Table 5: Treatment Options for Men with Erectile Dysfunction

Treatment	*Advantages*	*Disadvantages*	*Recommendation*
Psychosexual therapy	Noninvasive Partner involved Curative	Time-consuming Patient resistance	First line treatment May be combined with other treatments
Sildenafil	Oral dosage Effective	Cardiovascular disease is a contraindication in some men	First line treatment Contraindicated with nitrates
Transurethral alprostadil	Local therapy Few systemic side effects	Moderately effective (43–60%) Requires office training Causes penile pain	Second line treatment
Intracavernous alprostadil or drug mixtures	Highly effective (up to 90%) Few Systemic side effects	Requires injection High dropout rate Can cause priapism or fibrosis Causes penile pain	Second line treatment
Vacuum constriction device	Least expensive No systemic side effects	Unnatural erection Causes petechiae Causes numbness (2%) Trapped ejaculation	Second line treatment
Surgical treatment Prosthesis (all types)	Highly effective	Unnatural erection (semirigid device) Infection Requires replacement in 5–10 year Requires anesthesia and surgery	For men not satisfied with medical treatment
Vascular surgery	Curative	Poor results in older men with generalised disease Requires anesthesia and surgery	For young men with congenital or traumatic erectile dysfunction

Epidemiology. The exact prevalence of this disorder is not known. It constitutes about 30 to 60 per cent of patients presenting with a psychosexual problem. The patient is typically a young male; more likely to be married or recently married, of average or low socio-economic status; student, labourer or farmer by occupation; comes from a rural or semiurban area and belongs to a family with conservative attitudes towards sex. The other factors like literacy, religion etc., are unimportant.

Clinical Picture. The passage of a whitish discharge with urine is described as 'Dhat,' believed to be semen by the patient, although there is no objective evidence to it. Sometimes an overvalued idea that the semen has become foul-smelling or less viscous in consistency may be present. The patient perceives the slightest turbidity in urine with unbelievable perturbance.

The individuals with Dhat syndrome present with vague somatic symptoms (like fatigue, weakness, anxiety, loss of appetite, etc.), psychological symptoms (like guilt, sad mood, lack of concentration and memory etc.) and at times, sexual dysfunctions (impotence, premature ejaculation), which the patient attributes to the passing of semen (*Dhat*) in urine as a direct consequence of his excessive indulgence is masturbation or sexual intercourse.

In a study (Bhatia et al., 1989), the knowledge and attitudes of patients toward Dhat syndrome were studied. The patients believed Dhat as semen; sugar; concentrated urine or not known. Scientifically, Dhat has been reported as the whitish discharge which passes along or before the passage of urine. This is usually related to the presence of oxalate and/or phosphate crystals which are present in a high concentration in the alkaline urine of the average Indian consuming vegetarian diet.

A majority of patients believed masturbation or excessive sex as the major cause of Dhat syndrome and expressed the belief that it might lead to physical and mental weakness and require some medication.

Differential Diagnosis

i. ***Physical Illnesses.*** Dhat syndrome needs differentiation from a number of physical diseases (e.g., sexually transmitted diseases, diabetes mellitus, etc), drugs intake (antacids, tonics, etc.) and spermatorrhoea which may produce turbidity in urine.

ii. ***Psychological Disorders.*** The primary psychiatric illnesses e.g., Anxiety states, depression, somatoform disorder etc., if present needs appropriate simultaneous attention.

The careful detailed history and routine investigations (urine examination, VDRL etc.) may help in ruling out physical or psychological problems.

The main treatment of Dhat syndrome is education, counselling and reassurance. The use of antianxiety or antidepressant medication is required when the symptoms of anxiety or depression are prominent, but they should be used for a few weeks only and the use of dependence producing drugs should be avoided. The excessive intake of fluids and acidification of urine (by prescribing ascorbic acid) may help in removing the turbidity due to concentrated or alkaline urine. The so-called 'tonic' or injectable anabolic steroids should be avoided, as they themselves can produce turbidity in urine.

The associated psychosexual dysfunctions (impotence, premature ejaculation etc.) need appropriate intervention.

Outcome. The outcome of 'pure' Dhat syndrome is good. Most of the patients recover completely. However, a small number of patients, who fail to receive what they desire, may drop out and indulge in "Doctor-shopping".

Other disorders are listed in Table 6.

Paraphilias (Sexual Deviations)

The paraphilia disorders are characterised by repetitive or preferred sexual fantasies or acts that involve non-human objects or non-sonsenting partners. Table 7 summarises various paraphilias.

Gender Identity Disorders

Disturbances in gender identity is rare and should not be confused with the far more common phenomena of feelings of inadequacy in fulfilling the expectations associated with one's gender role e.g., an individual who perceives himself or herself as being sexually unattractive yet experiences himself or herself unambiguously as a man or woman in accordance with his or her anatomic sex.

i. Transsexualism (Gender Dysphoria Syndrome)

Individuals with this disorder usually complain that they are uncomfortable wearing the clothes of their own anatomic sex; frequently this discomfort leads to cross-dressing (dressing in clothes of the other sex). Often they choose to engage in activities that in our culture tend to be associated with the other sex. These individuals often find their genitals repugnant, which may lead to persistent requests for sex reassignment by surgical or hormonal means.

To varying degrees, the behaviour, dress and mannerisms are those of the other sex. With cross dressing, hormonal treatment, and electrolysis a few males with the disorder will appear relatively indistinguishable from members of the other sex. However, the anatomic sex of most males and females with the disorder is from members of the other sex. However, the anatomic sex of most males and females with the disorder is apparent to the alert observer. Generally there is moderate to severe coexisting personality disturbance. Frequently there is considerable anxiety and depression which the individual may attribute to inability to live in the role of the desired sex.

ii. Gender Identity Disorder of Childhood

Boys with this disorder invariably are preoccupied with female stereotypical activities. They may have a preference for dressing in girl's or women's clothes, or may select such items from available material when genuine articles are unavailable (the cross-dressing never causes sexual excitement). They often have a compelling desire to participate in the games and pastimes of girls. Dolls are often the favourite toy, and girls are regularly the preferred playmates. In rare cases, a boy with this disorder claims that his penis or testes are disgusting or will disappear. Girls with this disorder regularly have male peer groups, an avid interest in sports and rough-and-tumble play and a lack of

Table 6: Psychosexual Dysfunctions and their Management

	Diagnostic criteria (Persistant and recurrent)	*Aetiology*	*Management*
Hypoactive Sexual Desire Disorder	Deficient or absent sexual fantasies and desire for sexual activity depending age, sex and context of person's life.	Testosterone deficiency, chronic physical illness, depression, prolactinemia, drugs (beta blockers, sedatives) and psychological factors.	Testosterone replacement if there is deficiency. Management of contributing factor. Psychotherapy to overcome inhibition and learning stress management techniques. *Marital therapy* to determine and address interactive factors in the couple.
Sexual Aversion Disorder	Extreme aversion to and avoidance of all or almost all, genital sexual contact with a sexual partner.	Attitudes and conflicts during early development; Marital discord.	Individual Psychotherapy to reduce the patient's fear and avoidance of sex. Marital Therapy, Sex Therapy, Medications (Tricyclic anti-depressant or anti-anxiety drugs).
Female Arousal Disorder	Partial or complete failure to attain or maintain the lubrication-swelling response of sexual excitement until completion of the sexual act.	Psychological conflicts, lack of desire, Dyspareunia, estrogen deficiency, marital discord and medications (antihistaminics, anticholinergics).	Estrogen replacement therapy, Correction of any medical or physical cause, Psychotherapy and Marital therapy to correct psychological factors.
Female Orgasmic Disorder	Delay or absence of orgasm following a normal sexual excitement phase during sexual activity.	Cultural restrictions, financial concerns, extra-marital affair, alcohol abuse, physical illness, drugs (antidepressants) and other psychological factors.	Drugs (see end of chapter) Treat organic causes Behavioural techniques Sensate focus, anxiety management. Strengthening pubococcygeal muscles, experiencing orgasm with a partner. Group therapy, Individual psychotherapy, Marital therapy.

Contd...

Table 6: Psychosexual Dysfunctions and their Management (*Contd...*)

	Diagnostic criteria (Persistant and recurrent)	*Aetiology*	*Management*
Male Orgasmic Disorder	Delay or absence of ejaculation following an adequate phase of sexual excitement.	Drugs (e.g. α-adrenergic blockers) Neurological disorder, surgical procedures, psychological factors (fear of trauma).	Correct physical factors, Identification and desensitisation of Psychological factors.
Premature Ejaculation	Ejaculation with minimal sexual stimulation before, upon, or shortly after penetration and before the person wishes it (taking into account the age, novelty of partner or situation and frequency).	Anxiety, early sexual experiences, Guilt about sex or hostility towards the partner.	Behaviour techniques such as sensate focus, anxiety reduction, squeeze or stop and go technique. Psychotherapy to couples and individuals.

Table 7: Some Characteristics of Paraphilias (Sexual Deviations)

	Diagnostic criteria (Repeated, preferred or exclusive method of achieving sexual excitement by —	*Age at onset*	*Course*	*Differential diagnosis*
Fetishism	Use of non-living objects (fetishes) the fetishes are not limited to articles of female clothing or used for purpose of sexual stimulation.	Adolescence or childhood	Chronic	*Non pathological sexual experimentation* (Stimulus is neither preferred nor required) *Transvestism.* Sexual arousal limited to articles female clothing used in cross-dressing.
Transvestism	Recurrent and persistent cross-dressing by a heterosexual male for purpose of sexual excitement.	Childhood or early adolescence	Chronic may develop into transexualism	*Transexualism*: No sexual excitement with cross- and persistent wish to be rid of one's own genitals. *Female impersonators* act does not cause sexual arousal or interference does not cause frustration *Fetishism.* It is not diagnosed when limited to articles of female clothing.
Zoophilia	Act of fantasy of engaging in sexual activity with animals.	No information	Chronic	*Non pathological sexual activity* with animals when suitable human partners are not available.
Pedophilia	Act of fantasy of engaging in sexual activity with pre-pubertal children.	Adulthood or middle age	Unknown (Chronic)	*Isolated Sexual Acts* with children may be precipitated by marital discord, rental loss or intense loneliness. Mental retardation Organic Personality Syndrome, *alcohol intoxication* or *Schizophrenia* only isolated acts are there. *Exhibitionism* Exposure may be a child but the act is not prelude to further sexual activity *Sexual Sadism.*

Contd...

Table 7: Some Characteristics of Paraphilias (Sexual Deviations) (*Contd...*)

	Diagnostic criteria	*Age at onset*	*Course*	*Differential diagnosis*
Exhibitionism	Acts of exposing the genitals to an unsuspecting stranger with no attempt at further sexual activity with the stranger.	Pre-adolescence to middle age (commonest mid-puberty)	Chronic	*Repeated exposure* without experiencing sexual excitement. *Pedophilia* when exposure occurs, it is a prelude to sexual activity with child.
Voyeurism	Observing unsuspecting people who are naked, in sexual activity and no sexual activity with the observed people is sought.	Early adulthood	Chronic	*Normal sexual activity* It is not with an unsuspecting partner and is usually a prelude to further sexual activity *Watching pornography.* The people being observed are willingly in view.
Masochism	International participation in an activity in which one is humiliated, bound, beaten or otherwise made to suffer.	Variable (early adulthood)	Chronic	*Masochistic fantasies* Sexual masochism is diagnosed if the individual engages in acts *Masochistic Personality traits.* They are not associated with sexual excitement.
Sadism	Intentional infliction of psychological or physical suffering on a non-consenting partner.	Variable (early adulthood)	Chronic	*Rape or other sexual assault.* A rapist not motivated by the prospect of inflicting suffering and may even lose sexual desire as a consequence.

interest in playing with dolls. More rarely, a girl with this disorder claims that she will grow up to become a man (not merely in role) that she is biologically unable to become pregnant, that she will not develop breasts, or she has, or will grow, a penis.

Some children refuse to attend school because of teasing or pressure to dress in attire stereotypical of their sex. Most children with this disorder deny being disturbed by it except as it brings them into conflict with the expectations of their family or peers.

Some of these children, particularly girls, whom no other signs of psychopathology. Others may display serious signs of disturbance as phobias and persistent nightmares.

Peer relations with members of the same sex are absent or difficult to establish. The amount of impairment varies from none to extreme and is related to the degree of underlying psychopathology and the reaction of the peers and family to the individual's behaviour. In a small number of cases, the disorder becomes continuous with transsexualism.

Homosexuality

The essential features of ego-dystonic homosexuality are a desire to acquire or increase heterosexual arousal, so that heterosexual relationships can be intiated or maintained, and a sustained pattern of overt homosexual arousal that the individual explicitly states has been unwanted and a persistent source of distress.

Diagnosis. This category is reserved for those homosexuals for whom changing sexual orientations is a persistent concern and should be avoided in cases where the desire to change sexual orientations may be a brief temporary manifestation of an individual's difficulty in adjusting to a new awareness of his or her homosexual impulses. Sexual behaviour is highly variable, ranging from:

(*i*) Exclusive heterosexuality

(*ii*) Sporadic homosexual encounters, usually in adolescence and early adulthood

(*iii*) Bisexuality (attraction to partners of both sexes)

(*iv*) Exclusive homosexuality

The individuals with this disorder is prone to dysthymic disorder, depression, anxiety, guilt and other medical illness like AIDS, hepatitis and sexually transmitted diseases. Loneliness is particularly common. '*Homophobia*' refers to irrational fear of homosexuality and societal attitudes, including prejudice, discrimination and harassment are distressing for male and female homosexuals.

14

Common Childhood and Adolescent Disorders

Like adults, children may experience disturbance in emotions, behaviour and relationships which impairs their functioning. It is distressing to the child as well as parents and the community. Most problems encountered with young children can be broadly classified into problems of emotions or behaviour.

It becomes imperative to identify these problems early because of several reasons:

- Children would rarely seek help from professional agencies directly
- Symptoms if left unchecked may persist into personality problems, disturbed interpersonal relationships, poor academic achievement, low self esteem etc.

Judicious early identification would curtail needless suffering and avoid spiralling of problems.

There is no one cause for these disturbances. Reasons are often multiple: genetic, environmental, chromosomal and socio-cultural. Factors like child's temperament, parental health, family relationships and parenting styles are important.

Influences on Child's Behaviour

- ***Preconception factors*** i.e., age of parents, intensity of their desire for a child.
- ***Prenatal factors*** i.e., maternal diseases, psychological stress, preterm delivery
- ***Postnatal factors*** such as
  - Establishment of a bond between parents and a child.
  - Parents and the home i.e., love for the child, fear of spoiling, overprotection, favoritism as well as rejection, parental habits etc.
  - Disturbed family e.g. inadequate family, antisocial family.
  - Attitude of other significant person e.g. teachers, friends and siblings.

Child Temperament

Despite these environmental influences and stressors certain children are more vulnerable while some are less. Children differ in their personality character or temperament. A "difficult child". Three broad types of temperament were characterised: children who were regular, predictable, and showed generally positive reactions - the easy babies, those who were almost the opposite, whom Chess called the *'mother killers'*, and a sizeable minority who were 'slow to warm up' to new situations but who adjusted eventually. It was found that those who developed behavioural problems in later childhood were difficult temperament babies.

Comprehensive Evaluation

Comprehensive evaluation of the child should include:

- Clinical Interviews
- School Report
- Intellectual Functioning
- Development Tests
- Neurological Assessment

For this reason, children with problem are often seen in a multidisciplinary setting including inputs from pediatrician, psychologist, occupational therapist, special educator and speech therapist.

Described below are some of the common childhood disorders

ATTENTION DEFICIT HYPERACTIVITY DISORDER

All children are active, but few are extraordinarily so and are considered hyperactive. The physical movements of the child are excessive and beyond normal acceptable limits. The activity is inappropriate and undirected rather than purposeful or productive.

Excessive physical movement (beyond a normal or acceptable limit) is termed hyperactivity. Reasonably objective parents can recognise when the amount and degree of activity (constant and involuntary) is different than that of peers of the same sex. When in doubt, they can be suggested to visit a classroom or play area for children of the same age. Requesting a friend to observe and provide objective information regarding comparative activity levels is especially helpful. A very useful concept is that hyperactivity is indicated by the inappropriateness and un-directness of the activity, as compared to the very active but purposeful and productive child.

It must be noted that high activity levels are typical in children who are normal 2 and 3 year olds, mentally aged 2 or 3, highly exploratory and very intelligent, overly nagged by adults and environmentally deprived.

Some other features of such a child are

- Fidgetiness
- Difficulty in remaining seated when required

- Easy distractibility
- Difficulty in taking turns
- Blurting out answers even before question is completed
- Shifting from one incomplete activity to another
- Excessive talkativeness
- Poor listening skills
- Carelessness and tendency to loose things
- Engaging in dangerous activity with little consideration for possible consequences i.e., Impulsivity.

When parents have a child with any type of problem, it is not unusual for the parents to feel guilty or blame themselves. Parental education and support has been shown to be very helpful for families with an ADD/ADHD child and its value cannot be overemphasised. However, as scientists and medical professionals learn more about the possible causes of this condition, there is growing evidence that parents likely had little control over the cause of ADD/ADHD. The following are commonly accepted as the most likely causes:

Causes

The exact causes of ADHD are unknown. Some suggested factors include:

1. Minimal and subtle brain damage during fetal development and early infancy. This may be caused by infection, mechanical damage, pre-maturity and so on.
2. Pregnancy/birth complications: Premature birth, lack of oxygen, or history of prenatal exposure to drugs/alcohol.
3. Heredity/Genetic predisposition: Another member of the family—grandparent, uncle, aunt, etc. had a similar temperament or pattern of behaviour. If one person in the family is diagnosed as ADHD there is greater probability of ADHD in another direct blood relation which suggests a genetic basis to ADHD.
4. Food additives, colouring agents, preservatives and sugar.
5. Prolonged emotional deprivation, stressful life events may initiate or perpetuate ADHD.
6. Biological/physiological causes: Possible chemical imbalance that inhibits the efficiency of the neurotransmitters of certain portions of the brain.
7. Lead poisoning: Ingesting toxic levels of lead, either by mouth or absorption.
8. Allergic/medical conditions: Predisposition toward asthma, food allergies and ear infections.

The issue of hyperactivity and inattention is further complicated by the fact that the same child who can't sit still to finish dinner may be able to attend and persist in activity of his interest, e.g., movie watching for 3 hours at a stretch.

If properly treated, most children with ADHD can live productive lives and can cope reasonably well with their symptoms.

Management

(a) Pharmacological

(i) Stimulants reduce symptoms in about 75%; they improve self-esteem by improving the patient's rapport with parents and teachers. Stimulants decrease hyperactivity. Plasma levels are not useful.
   (a) Dextroamphetamine is approved by the Food and Drug Administration (FDA) for ages 3 years and over.
   (b) Methylphenidate is FDA approved for ages 6 years and older. The sustained-release preparation does not have proven usefulness.

(ii) Clonidine and Guanfacine are reported to reduce arousal in children with the disorder.

(iii) Antidepressants if stimulants fail; may be best in ADHD with symptoms of depression or anxiety. Imipramine and Despiramine have shown some efficacy in studies, but safety of the medication is suspect. Buspirone/Bupropion are also useful.

(iv) Antipsychotics or Lithium if other medications fail but only with severe symptoms and aggression (concomitant disruptive behaviour disorder).

(b) Psychological—Multimodality treatment is necessary for child and family. It may include medication, individual psychotherapy, family therapy, and special education (especially with coexisting specific developmental disorder). These interventions are crucial in moderate or severe cases, given the risk of delinquency.

CONDUCT DISORDER

Prevalence ranges from 5–15% in studies. Conduct disorder accounts for many impatient admissions in urban areas. The male-to-female ratio is 4–12:1.

Clinical Features

A repetitive and persistent pattern of behaviour in which the basic rights of others or major age-appropriate societal norms or rules are violated, as manifested by the presence of three (or more) of the following criteria in the past 12 months, with at least one criterion present in the past 6 months.

Aggression to people and animals

1. often bullies, threatens, or intimidates others
2. often initiates physical fights
3. has used a weapon that can cause serious physical harm to others (e.g., a bat, brick, broken bottle, knife, gun)
4. has been physically cruel to people
5. has been physically cruel to animals
6. has stolen while confronting a victim (e.g., mugging, purse snatching, extortion, armed robbery)
7. has forced someone into sexual activity

Destruction of property

8. has deliberately engaged in fire setting with the intention of causing serious damage
9. has deliberately destroyed others' property (other than by fire setting)

Deceitfulness or theft

10. has broken into someone else's house, building, or car
11. often lies to obtain goods or favours or to avoid obligations (i.e., "cons" others)
12. has stolen items of nontrivial value without confronting a victim (e.g., shoplifting, but without breaking and entering; forgery)

Serious violations of rules

13. often stays out at night despite parental prohibitions, beginning before age 13 years
14. has run away from home overnight at least twice while living in parental or parental surrogate home (or once without returning for a lengthy period)
15. is often truant from school, beginning before age 13 years

General Considerations

Conduct disorder is associated with family instability, including victimisation by physical or sexual abuse. Propensity for violence correlates with child abuse, family violence, alcoholism, and signs of severe psychopathology, e.g., paranoia and cognitive or subtle neurological deficits. It is crucial to explore for these signs; findings can guide treatment.

Conduct disorder often coexists with ADHD and learning or communication disorders. Suicidal thoughts and acts and alcohol and drug abuse correlate with conduct disorder.

Some children with conduct disorder have low plasma dopamine β-hydroxylase levels, abnormal serotonin levels.

Management

(a) ***Pharmacological*** – Lithium or haloperidol is of proven efficacy in many aggressive children with conduct disorder. Carbamazepine has shown success. β-Adrenergic receptor antagonists deserve study. SSRIs are useful for reducing impulsivity and aggressive behaviour.

(b) ***Psychological*** – Multimodality as in ADHD. It often includes medication, individual or family therapy, tutoring, or special class placement (for cognitive or conduct problems). It is crucial to discover and fortify interests/talents to build resistance to the lure of crime. If environment is noxious or if conduct disorder is severe, placement away from home may be indicated.

THE DEPRESSED CHILD

Only recently has there been a growing awareness of the increasing number of depressed children under 12 years. Currently, it is estimated that 1 in 5 children have some form of depression.

Clinical Features

Depressed children rarely show joy or pleasure, often have a soft monotonous voice, lack of sense of humour, and rarely laugh. There may be mood swings and disturbed sleep patterns. They may be tearful, irritable, miserable, and cling for support. Some become detached and aloof, while others appear overtly anxious. Rather than complaining of sadness, children may have physical complaints (headaches, stomach aches). They may not feel like doing anything, lose interest in playing sports or games, and suddenly do poorly in school. It may be hard to arouse their interest in anything. They may feel rejected and unloved and not be easily comforted. Often, they prefer isolated, self-comforting activities to interacting with others. To others, they may seem too serious, solemn, and grown up.

General Considerations

Self-injurious behaviour occurs when people damage or hurt themselves. Suicide is an extreme form of self-injurious behaviour that often occurs in depressed individuals.

Strikingly, 58 per cent of parents of depressed children are also depressed. To make the situation even more serious, there is a widespread professional opinion that depression at times underlies a variety of children's behaviour problems such as bedwetting, tantrums, truancy, fatigue, school failure, delinquent acts, hyperactivity, and psychosomatic problems.

Delinquency often covers up feelings of loneliness and despair. Approximately half of the adolescents in trouble with the law are depressed. This situation is called a *"masked depression."*

Common Treatment Approaches

A large number of treatment strategies have been developed for the treatment of depression. Many of these approaches can be implemented individually, in groups or family therapy environment. There is considerable evidence to suggest that interventions which emphasise treatment of the family, and not the "identified patient," are critical to positive treatment outcome. Peer group approaches have been found to be effective for children. Play therapy is sometimes appropriate with younger children.

(a) **Psychological**

- **Cognitive:** Cognitive approaches utilise specific strategies that are designed to alter negatively based cognitions. Depressed patients are trained to recognise the connections between their thoughts, feelings, and behaviour; to monitor their negative thoughts; to challenge their negative thoughts with evidence; to substitute more reality-based interpretations for their usual interpretations; and to focus on new behaviours outside treatment.
- **Behavioural:** Behavioural approaches designed to increase pleasant activities include several components such as self-monitoring of activities and mood, identifying positively reinforcing activities that are associated with positive feelings, increasing positive activities, and decreasing negative activities.

- **Social skills:** Social skills training consists of teaching children how to engage in several concrete behaviours with others. Initiating conversations, responding to others, refusing requests, making requests, etc. Patients are provided with instructions, modelling by an individual or peer group, opportunities for role playing and feedback. The object of this approach is to provide the child with an ability to obtain reinforcement from others.
- **Self-control:** Self-control approaches are designed to provide the self-control strategies including self-monitoring, self-evaluation and self-reinforcement. Depressive symptoms are considered to be the result of deficits from one or more areas and are reflected in attending to negative events, setting unreasonable self-evaluation criteria for performance, setting unrealistic expectations, providing insufficient reinforcement, and too much self-punishment.
- **Interpersonal:** Interpersonal approaches focus on relationships, social adjustment and mastery of social roles. Treatment usually includes non-judgmental exploration of feelings, elicitation and active questioning on the part of the therapist, reflective listening, development of insight, exploration and discussion of emotionally laden issues, and direct advice.

(b) **Pharmacological**

Several classes of medications are used with adult populations. Major types include, tricyclics (e.g., Imipramine and Amitriptyline), SSRIs, (e.g., Fluoxetine, Paroxetine, Sertraline, Fluvoxamine, Citalopram, Escitalopram) and Monoamine Oxidase Inhibitors (e.g. Phenelzine) but other classes have emerged as well. These drugs are not without side effects. These drugs have been shown to be 50–70% more effective with adults than placebos and no other treatment. Very little is known about the safe use of antidepressants with children. SSRIs as a class of drugs in children is safer. The risks and side effects of medications and the findings that competent therapy and counselling interventions may be more effective restrict the use of medications with children.

Some common features of depressed children: What help can be given?

* Guilt • Anger Turned Inward • Feeling Helpless • Gain Attention, Love, Sympathy or Revenge • Reaction to Tension • Family Context

* Open Communication and Expression of Feelings • Promote Adequacy and Effectiveness • Promote Many Sources of Self-esteem • Model Optimism and Flexibility • Be Alert to Warning Signs.

THE BEDWETTING CHILD

Clinical Features

Bedwetting or enuresis can be defined as the repeated, involuntary discharge of urine into the bed by a child age 4 or older. An occasional wet bed is not considered problem; most bedwetters wet several nights a week or every night.

Two types of bedwetters have been identified in the literature: the **primary bedwetter** who has been wetting since birth, and the **secondary wetter** who had achieved a significant period of night time dryness (at least 3 months) and then resumed bedwetting.

Causes

There are many theories as to the underlying causes of bed wetting, but none have been conclusively proven.

***Primary bedwetter*:** Some external stress or emotional crisis that makes the child anxious, such as the birth of a new sibling, physical illness, or a family move.

***Secondary bedwetter*:** The most plausible explanation for bed wetting is a maturational lag, i.e., slow physiological maturation of bladder control mechanisms. This seems to be inherited since the parents of enuretic children have a history of being enuretic themselves.

Others suggest that bed wetting is the result of a development lag interacting with maladaptive toilet training practices.

General Consideration

In toilet training a child, avoid the extremes of being lax and indifferent about bladder control and being overly punitive, i.e., severely punishing, scolding, or shaming the child for wetting the bed. Punitive methods tend to make the child feel guilty, inadequate, and/or anxious. When a child is markedly anxious or fearful, it is hard for him to learn new behaviours, such as night time control. It is also wise to delay bladder training until the child is comfortable about daytime control, has bowel control, and is able to hold urine for several hours at a time.

It has been suggested that if parents pressure a child to achieve night time control before the child is mature enough, the child may lose confidence and have greater difficulty in bladder control. Some investigators feel that if parents were to completely ignore the slow development of bladder control in the child, this would lead to spontaneous cure by the time the child reaches 7 to 8 years. Unfortunately most parents become quite upset about the bedwetting so that the child becomes anxious and discouraged which makes the problem worse.

Some parents become emotionally cold and distant to the child, while others try to reason with bedwetters to get them to "try harder." In general, these tactics are not only ineffective but may actually make the problem worse. Above all, parents are best advised to respond to the bedwetting in a calm, matter-of-fact manner and show confidence in the child's ability to eventually control this behaviour.

Management

(a) ***Pharmacological***—The use of drugs, such as tricyclic antidepressants (Imipramine), has been found to alleviate the bed wetting in one out of every three bedwetters. In using any procedure, it is important to obtain the child's cooperation and make it a joint effort to overcome the problem, rather than attempting to force a method on a child. Desmopressin, nasal spray, and tablet oxybutynin are also useful.

(b) ***Psychological***—Some limited success has been achieved by limiting the amount of fluid a child can drink after 6 p.m. and by requiring the child to urinate before going to bed.

***Star Chart*:** Ask the child to keep a record of wet and dry nights. Dry nights are highlighted on the chart with gold stars, and a reward, such as extra time alone with a parent, is given for each stage of improvement. While ignoring wet nights, the parents praise the child for each dry night. Such a reward system helps motivate the child by giving him an incentive, a specific goal to achieve, and a picture of ' is progress in reducing the habit. Star charts have been found to be particularly effective with young bedwetters.

***Reduce Stress*:** If the child had been dry at night and then started wetting again, check to see if some stress occurred just prior to the resumption of bed wetting, such as the birth of a new sibling, a move to a new neighbourhood, a family quarrel, or the extended absence of a parent for any reason. If an uncontrollable external stress seems to be triggering the bedwetting, do what you can to reduce the child's anxiety by giving extra attention, support, and understanding to the child.

At bed time sit with the child for 10–15 minutes of comforting talk to the child goes to sleep relaxed, feeling sure of your interest and support. You might also spend extra time during the day interacting on a one-to-one basis with the child in pleasant activities. Use this time to observe the child and to try to uncover any areas of conflict or unresolved anxiety.

Some parents require the school-age bedwetter to change the bed sheets after wetting and to see that the bed clothes get washed. This seems to be a logical consequence to the act of wetting. Do not scold or lecture when establishing this procedure.

***Night time Awakening*:** The first step in this procedure is to determine what time the child usually wets the bed each night. If, for example, you find that the child usually wets 2 hours after retiring, set an alarm clock to go off in the child's room just before this time. When the alarm goes off, the child arises, urinates in the toilet, and sleeps through until morning. After 7 consecutive dry nights is the criterion for further gradual reductions in time, to 60 minutes after bed time, to 45 minutes, and finally to 30 minutes. The child is then to go to the toilet every other night without the alarm clock to fade out its use. When this procedure was used with a 13 years old girl who also cleaned her bed sheets when wet, the bedwetting was soon eliminated.

AUTISM

Autism is developmental disability marked by significant impairments in social relatedness, communication, and the quality, variety, and frequency of various activities and behaviours. The onset of autism generally is before age 3, and impairment persists throughout the lifespan. Autism may occur across a range of functioning, and is often associated with mental retardation. Median prevalence estimate is 4–5 per 10,000 M : F ratio is 3 : 5 or 4 : 1.

Clinical Features

A constellation of symptoms is always seen in autism spectrum disorders, and a variety of other symptoms are often, although not always associated with the disorder. The presentation of these symptoms varies greatly among individuals, and no single symptom is pathognomonic.

Primary Symptoms Include the Following:

Abnormal Social Relatedness

Social relatedness is always impaired in autism. The degree of impairment, however, may range from an oddness in social interaction to an almost complete detachment and lack of responsiveness to other's social initiations. Social abnormalities may include poor use of eye contact, emotional cues, and social smile; lack of social initiation and disorganised patterns of reactions to strangers and separations. Children with autism demonstrate a particular inability to imitate others in typical ways, with disregarding the other or, occasionally, inappropriately mirroring the other's behaviour.

Abnormal Communicative Development

Much of the literature on autism has focused on deviance in the development spoken language. However, the communication deficit is much more profound than impaired language alone.

Abnormal Capacity for Symbolic Play

Children with autism are particularly lacking in the pretend play typical of preschool-aged children, including doll play, role play and dramatic play, individuals with autism rarely seek out play partners.

Restricted and Odd Behavioural Repertoire

Typical toy play, marked by curiosity, exploration interest in novelty, and goal directedness is lacking in children with autism. Much time is spent in a very limited range of activities, which may consist of a few highly ritualised or repetitive ways of handling a few object (e.g. sucking, shaking arranging, carrying around). When age-appropriate play skills are present, they are often inappropriately repetitive. Water play, watching things move or spin, and watching television commercials or videos are typical interests of young children with autism.

Extreme Behavioural Problems

A minority of children with autism exhibit extreme behavioural difficulties such as self abuse, high levels of aggression and destruction, and difficult to manage behaviour.

Pervasive Developmental Disorder

Infantile autism, childhood schizophrenia. Some other conditions such as disintegrative psychosis, Asperger's syndrome and Rett's syndrome are also included described with autistic features.

Management

The goal of treatment is to reduce disruptive behaviour and to promote learning particularly language acquisition and communication and self-help skills.

*(a) **Pharmacological***

No pharmacological agent has proved curative but certain medications may be of benefit for specific symptoms such as self-injury, aggression, stereotyped movements and over activity. Haloperidol and Risperidone: May decrease stereotyped behaviours and agitation. Fluoxetine and Citalopram: Reduce repetitive behaviour and impulsive aggressions.

(*b*) ***Psychological***

Autism is generally considered a lifelong, chronic disability. Nevertheless, specific educational and therapeutic interventions are critical for stimulating development in all areas and improving the person's adaptive functioning in all settings (home, school, work and community).

Adolescents and adults with autism frequently need specific help in negotiating the complexities of life demands. Social skills groups, recreational activities, individual psychotherapy, and vocational coaching and assistance can help them acquire skills necessary for a satisfying adult life.

With appropriate educational and treatment services, children with autism will show some improvements. The preschool years are typically the most difficult, because children with autism tend to be the least social, least communicative, and have the most difficulties behaviourally.

IQ remains stable across the lifespan, but the severity of the social and communicative deficits tends to diminish as children grow older. Learning continues throughout childhood and adolescence, as long as children are receiving appropriate services. Adolescence can be a difficult time for some individuals with autism, because of increased sexual behaviour and aggressiveness.

The most important positive prognostic indicators are functional language before age 5, and cognitive abilities above the mentally retarded range (i.e., IQ > 70). Another 30% are reported as achieving some degree of partial independence in adulthood. Finally, roughly 25% of children with autism develop seizures beginning in adolescence or early adulthood.

MENTAL RETARDATION

In mental retardation the individual operates at a level significantly below the intellectual functioning of the general population, resulting in difficulties of problem solving the adaptation over a wide area of functioning.

Intellectual capacity is an essential element of daily living and normal scholastic and social problem solving skills. In profound or severe retardation, the slower rate of learning skills rather than absolute inability marks out the child.

Judgement of mental retardation must be based on rigorous intellectual assessment, using relevant and valid test, carefully interpreted to allow for the child's physical, emotional, motivational and social features. The varying degrees of mental retardation should be identified based on comprehensive ability of the child and not upon any specific deficiency.

Discriminating Features

1. Significantly sub average intellectual functioning.
2. Concurrent deficits or impairments in present adaptive functioning.

Consistent Features

1. An IQ of approximately 70 or below on an individually administered IQ test.
2. Impairments in adaptive functioning in at least two of the following skill areas: communication, self-care, home living, social-interpersonal skills, use of community resources, self-direction, functional academic skills, work, leisure, health and safety.
3. Onset before age 18 years.

Variable Features

1. Prevalence is higher in males.
2. Stereotypies and self-injurious behaviours are often found in the moderate to severe forms of MR.
3. Often accompanied by other psychiatric disorders.

Management

Many behavioural interventions are aimed at providing alternatives to unwanted behaviours for patients with mental retardation. The rationale is that individuals acquire more appropriate means of obtaining desired ends; the frequency of aberrant behaviours may be reduced. These techniques are quite successful in individuals with mild and moderate impairment.

Conclusion

The psychiatric assessment of children and adolescents is a complex process. However, it can be made less intimidating when a basic framework is followed. The components of the examination include meeting with the parents and the child to gather a complete history, obtaining the mental status examination of the child, developing a treatment plan and presenting the results of the evaluation to the family. Yet, within this framework it is important to recognise individual differences between children and parents. The basic issue that is addressed is: whether the child has a problem that significantly interferes with his normal development, whether there are problems in social, academic or family functioning and whether there is an effective treatment that will allow the child to reach his potential free of interference from psychiatric morbidity.

Given the shortage of mental health professionals, collaboration with the general practitioners and family physicians becomes imperative in this regard.

Useful Child Psychiatry Web Sites

About our Kids	www.aboutourkids.org
Advocates 4 Special Kids	www.a4sk.org <http://www.a4sk.org>
American Academy of Pediatrics	www.aap.org <http://www.aap.org>
American Medical Association	www.ama-assn.org <http://www.ama-assn.org>
American Psychiatric Association	www.psych.org <http://www.psych.org>
American Psychological Association	www.apa.org <http://www.apa.org>

Contd...

American Society of Adolescent Psychiatry	www.adolpsych.org <http://www.adolpsych.org>
Autism Society of America	www.autims-society.org <http://www.autims-society.org>
Canadian Mental Health Association	www.cmha.ca <http://www.cmha.ca>
Canadian Paediatric Society	www.cps.ca <http://www.cps.ca>
Canadian Psychiatric Association	www.cpa-apc.org <http://www.cpa-apc.org>
Caring for Kids	www.caringforkids.cps.ca <http://www.caringforkids.cps.ca>
Centre for Mental Health Services	www.mentalhealth.org <http://www.mentalhealth.org>
Child Advocate	www.childadvocate.net <http://www.childadvocate.net>
Child and Adolescent Bipolar Foundation	www.bpkids.org <http://www.bpkids.org>
Children and Adults with Attention Deficit/Hyperactivity Disorder	www.chadd.org <http://www.chadd.org>

15

Disorder Related to Women

The psychiatric disorders ***more commonly reported in females*** include major depression, neurotic depression, anxiety state, phobic neurosis, hypochondriasis, adjustment problems, attempted suicide, anorexia nervosa and senile dementia.

The disorders which are peculiar to females are:

I. Premenstrual Syndromes

Menstruation (Menses) is a cyclic phenomenon usually occurring every 21 to 30 days and includes uterine bleeding for about 3 to 7 days.

Clinical Picture

— *Affective (Mood)*: Sadness, anxiety, anger, irritability, labile mood.

— *Cognitive*: Decreased concentration, indecision, suspiciousness, sensitivity, suicidal or homicidal ideation.

— *Pain*: Headache, breast tenderness, joint and muscle pain.

— *Psychological*: Insomnia, Hypersomnia, anorexia, craving for certain foods, fatigue, lethargy, agitation, libido change, hysteria and even psychosis.

— *Physical*: Nausea, vomiting, diarrhoea, sweating, palpitations, weight gain, swelling over legs, bloating, oliguria.

— *Dermatological*: Acne, dry hair, rash.

— *Neurological*: Fits, vertigo, dizziness, tremors, numbness, clumsiness.

— *Behavioural*: Decreased motivation, impulsivity, decreased efficiency, social withdrawal.

Epidemiology

Normal menstruation causes tension (Premenstrual syndrome) in about one-third to two-third of women.

The syndrome of 'pre, peri or paramenstrual tension' typically starts about 5 to 10 days before onset of menses and lasts till the end of menses. Premenstrual syndrome is also called ***Late luteal phase dysphoric syndrome.***

Etiology

(i) *Ovarian*: altered ovarian activity i.e., altered ratio of estrogen and progesterone.

(ii) *Fluid and Electrolyte (hormonal)*: Increased water and electrolytes retention, due to aldosterone rise in midcycle.

(iii) *Other hormonal*: Vitamin B or Magnesium deficiency: Changes in glucose levels; premenstrual change in endorphins; increased melatonin.

(iv) *Psychological*: Psychological e.g. anxiety neurosis, prolonged or excessive stress e.g. examination, marriage, divorce, marital disharmony, death or separation of a parent, depression, hysterical, inadequate or obsessive personalities etc. can precipitate premenstrual syndromes in 50–75 per cent of susceptible women.

Diagnosis

In 1968, ***Moos*** devised a 47 items "*Menstrual Distress Questionnaire*" consisting of 8 symptom groups. (see clinical picture above).

Treatment

The different kinds of treatment available for this disorder are:

(*i*) *Hormones*: Oral contraceptives, progesterone tablets, androgens, antihormonal (danazol) etc.

(*ii*) *Psychotropic Drugs*: Sedatives, antidepressants, lithium.

(*iii*) *Other Agents*: Pyridoxine, Vitamin B-complex, dietary restriction of salt etc., diuretics, bromocriptine, Prostaglandin inhibitor-analgesics (Aspirin etc.).

(*iv*) *Psychotherapy*: The treatment of associated physical, mood, behavioural or cognitive disturbances should also be treated with psychotherapy for better outcome.

II. Psychiatric Disorders of Childbirth

(a) Psychological Problems in Pregnancy

Epidemiology

Minor psychological symptoms are common, however, 66% of women have some psychological symptoms during pregnancy, especially in the first and last trimesters.

Clinical Picture

Anxiety is common, as is a tendency to irritability and minor lability of mood. About 10% pregnant women have depression more commonly in first trimester which usually lasts less than 12 weeks. It is often associated with previous history of abortion or depression, pregnancy being unwanted, marital complications and anxieties about the fetus. It is characterised by fatigue, irritability, increased neuroticism scores and denial of the pregnancy.

Etiology

The following factors may be associated

(i) Age of the mother
(ii) Parity
(iii) Relationship with her husband
(iv) Type of family
(v) Employment
(vi) Physical or mental complications during pregnancy
(vii) Attitude towards pregnancy
(viii) Changes during pregnancy
(ix) Other factors e.g., Immature personality, hostility towards husband and parents, unusual fears about self, about body changes, operation, job, the husband's love and affection.

Management

(i) *Counselling*: Increased support by medical, nursing and other services as well as by family and reduction in the need to contact psychiatric services.
(ii) *Medication*: Conjoint marital therapy or individual counselling of the husband may be used. Minor tranquilizers and tricyclic antidepressants may be indicated in second and third trimesters.

(b) Disorders of the Puerperium

(i) ***Normal postpartum reaction.*** First few days following delivery are psychologically stressful. Women frequently complain of anxiety, irritability, dysphoria, emotional stability, tearfulness, fatigue and mild vegetative symptoms (such as disturbances in appetite, sleep) and a desire for intimacy.
(ii) ***Puerperal disorders***

(c) Transitory Mood Disturbances (Postnatal Blues)

At least 50% women have a shortlived emotional disturbance commencing on the third day and lasting for 1 to 2 days. These are more common in primigravida and in those who complain of premenstrual tension.

Clinical Picture

Unfamiliar episodes of crying, irritability, depression, emotional lability, feeling separate and distant from the baby, insomnia and poor concentration. This coincides with sudden weight loss, decreased thirst and increased urinary sodium excretion.

Etiology

(i) *Biochemical*: The cause of postnatal blues is unknown but increased levels of urinary cyclic AMP and reduced plasma levels of free tryptophan have been reported.

(ii) *Others*: There is evidence that depression and mood instability are maximal on the fifth postpartum day and that women with higher neuroticism scores are more likely to experience 'the blues'.

(d) **Puerperal Neuroses**

Postnatal depression is not only the ***most frequent*** but also the most disabling neurotic disorder at this time.

Epidemiology

About 10 to 15% of mothers may develop a non-psychotic depression.

Onset is usually within the first postpartum month, often on returning home and usually between day 3 and day 14.

Predispositions

Postnatal depression is associated with *increasing age,* childhood *separation from* father; *problems* in relationship with mother-in-law and father-in-law; marital *conflict; mixed feelings* about the baby; *physical problems* in the pregnancy and prenatal period, a tendency to be more *neurotic* and less extrovert personalities; a *previous* psychiatric history; *family distress; lower* social class or a *hereditary* pre-disposition.

Clinical Picture

(See Table 1)

- loneliness or worry about a physical illness
- excessive anxiety about her baby's health that cannot be diminished by reassurance
- self-blame
- sad mood
- worry at her rejection of the baby
- irritability and loss of libido
- sleep difficulty
- a fear that baby may not be hers
- suicidal thoughts or a fear of harming the baby
- the other symptoms include feeling tired, despondent and anxious, poor appetite, decreased libido.

Management

The treatment consists of counselling and antidepressant drugs. Breast-feeding is not contraindicated but should be discontinued if treatment with lithium carbonate is maintained.

(e) Other Neurotic Disorders

Phobias, Anxiety states and obsessive-compulsive neuroses may also occur and interfere markedly with care.

(f) Puerperal Psychoses

Epidemiology

1 to 2 per 1,000 deliveries.

Clinical Picture

(See Table 1)

— Puerperal psychoses are not widely held to be distinct and unitary form of psychosis but to be divided into affective psychoses (70%), schizophrenia (25%) and organic psychoses (2 to 5%). In India and other developing countries, schizophrenic and organic types of psychosis are believed to be the more common than affective type.

Common symptoms of a puerperal psychosis are:

— severe insomnia and early morning waking.
— lability of mood, sudden tearfulness or inappropriate laughter.
— persistent perplexity, disorientation or depersonalisation.
— abnormal (unusual) behaviour such as restlessness, excitement or sudden withdrawal.

Table 1: Spectrum of Mood Disturbances in the Puerperium

	Postnatal blues	*Puerperal neuroses*	*Puerperal psychoses*
Frequency	50%	15% (13% depression)	0.2% Affective psychoses common; schizophrenic and organic psychoses rare
Peak time of onset	4–5 days after childbirth	2–4 weeks after childbirth	1–3 weeks after childbirth
Duration	Usually 2–3 days	4–6 weeks if treated; up to 1 year if not	6–12 weeks
	Severe 'blues' may→	Postnatal depression→	Affective psychosis
Profession of first contact	Midwife (hospital); obstetrician	Midwife (community); health visitor; general practitioner	Midwife; health visitor; general practitioner; psychiatrist (rarely)
Psychiatric referral	Virtually never	Unusual	Common—especially if marked behaviour disturbance
Possible treatments	Nil; but observation if severe	Counselling; antidepre-ssants	Admit to mother and baby unit; neuroleptics; antidepressants; ECT; lithium; counselling and advice about further pregnancy

— paranoid ideas that may involve close family relations or hospital staff.
— unexpected rejection of the baby or a conviction that baby is deformed or dead.
— suicidal or infanticidal threats.
— excessive guilt, depression or anxiety.

Etiology

The exact etiology is unknown but the following factors are important:

(*i*) *Genetic factors*: A family history of major psychiatric disorder; a past history of bipolar disorder also predisposes to the development of postpartum mood disorder.

(*ii*) *Biochemical factors*: Alterations in the hormonal levels of hypothalamic-pituitary-gonadal axis. The greater predelivery estrogen level (greater irritability) and lower postpartum estrogen level (sleep disturbance) and progesterone level (depression) are associated with symptoms of postpartum psychosis.

— larger increases in camp during pregnancy.
— urinary free cortisol excretion increases late in pregnancy, surges at birth and then rapidly declines. persistent increase in alpha-2-adrenoceptor capacity.
— alterations in thyroid hormone level.
— decreased endorphins levels are correlated with dysphoria, decreased motor activity, lability and lethargy.
— *Sleep.* Decrease in stage 4 sleep time and is correlated with mood irritability.

(*iii*) *Psychodynamic factors*: 'Patients' relationship with her own mother, her feelings about the responsibility of motherhood, her reaction to assertion of her female role, her relationship with her husband and his personality (overpassive or overdominant) and obsessive compulsive traits.

(*iv*) *Obstetrical factors*: Obstetrical events other than *parity* have not proven to be significant.

Management

(*i*) *Hospitalisation*: Puerperal psychosis is a psychiatric emergency.
Admission of both mother and baby together is always advisable if possible.

(*ii*) *Drugs and physical treatment* should be given as appropriate to the symptoms. Electroconvulsive therapy is very effective. If baby is breast fed, major tranquilizers may cause oversedation in baby.

(*iii*) *Psychotherapy* usually of supportive kind is required.

Outcome

About 70% recover fully, affective psychosis having a better prognosis than schizophrenic.

***Poor* prognosis is indicated by**

— a positive family history

— a history of schizophrenia
— neurotic personality
— presence of severe marital problems
— schizophrenic type of puerperal psychosis.

III. Termination of Pregnancy (Abortion)

Serious psychiatric illness is very rare following termination, the incidence being about 0.1 to 0.3 per 1,000 terminations.

Clinical Picture

The common psychological symptoms include—feeling of guilt and regret, emotional instability, a changed attitude towards sex, irritability, anxiety, depression, suicide, homicide, multiple somatic complaints (aches and pains), hysterical conversion symptoms and social and occupational maladjustment.

Since most of the psychological reactions following an abortion are short lasting and tend to disappear themselves in a few weeks to few months, they are appropriately called *'Post-abortion blues.'*

Management

Counselling, reassurance and supportive psychotherapy.

IV. Menopause

'Menopause' refers to the time of cessation of menstrual periods and can therefore only be noted in retrospect. The word *'climacteric'*s is defined as a critical phase in life when a major change is occurring but menopause is now also used with this wider meaning.

Clinical Picture

Table 2: Menopausal Index (Kupperman et al., 1959)

* Vasomotor symptoms	* Paraesthesia	* Insomnia
* Nervousness	* Melancholia	* Vertigo
* Weakness	* Arthralgia and Myalgia	* Headaches
* Palpitations	* Formication	

(*i*) ***Gynaecological view***: The various symptoms like depression, irritability, lack of confidence, poor concentration, autonomic symptoms are attributed to menopause.

The term ***'Menopausal syndrome'*** has been used to describe symptoms related to estrogen deficiency and include—hot flushes, sweats, atrophic vaginitis, osteoporosis and other symptoms depending on personality.

(*ii*) ***Psychiatric view*:** The belief that the menopause is a item of high risk for psychiatric disorder in women is not upheld in the psychiatric literature.

Etiology

(*i*) *Biological factors*: Abnormalities in

- — control of cortisol secretion.
- — response of thyroid stimulating hormone (TSH) to Thyrotropin releasing hormone (TRH).
- — response of growth hormone to clonidine.
- — fluctuations in level of prolactin, cortisols, TSH and Triiodothyronine.
- — upward LSH, FSH and Oestradiol.

(*ii*) *Psychoanalytic views*:

- — a time of great loss of femininity and reproductive potential.
- — a time of increased importance of penis envy.
- — loss of femininity and fear of growing old and associated loss of self-esteem.

(*iii*) *Cultural, social and family factors*: Negative expectations of menopause may be culture bound.

The major influences on the risk of developing depression are reported as—worries about work, adolescent children, ailing husbands and ageing parents.

Management

(*i*) *Hormone replacement*: Oestrogen therapy is believed to control many symptoms especially hot flushes and vaginal atrophy or dryness but the symptoms such as insomnia, irritability, palpitations, depression, vertigo, backache, fatigue and reduced libido are not relieved which respond to *other medication* such as antidepressants and benzodiazepines.

Combined oestrogen and testosterone implants may improve sexual problems especially loss of libido.

(*ii*) *Psychotherapy*: Explanation and reassurance.

16

Geriatric Psychiatry

Senescence is the normal process of growing old while ***Senility*** refers to the abnormal mental state which sometimes supervene towards the close of old life. The study of the physical and psychological changes which are incident to old age is known as ***Gerontology*** while ***Geriatrics*** is the study of the causes and medical treatment of ill health associated with old age.

POPULATION OF ELDERLY

In USA and England, about 12 to 14 per cent of the people are above 65 years of age, as against 3.8 per cent in India. According to 2001 Indian census, 7.0 per cent of the total population are above the age of 60 years.

Epidemiology

The estimates of psychiatric disorders in the elderly are based on some Western studies. It is estimated that about 50 to 60 per cent of elderly have psychiatric disorders. The exact prevalence of these disorders in India is not known but major depression is believed to be the ***commonest*** disorder.

The disorders which are common in old age are:

I. ORGANIC DISORDERS

(a) Transient Cognitive Disorder (Delirium, acute confusional state)

Delirium is defined as an organic brain syndrome characterised by global cognitive impairment of abrupt onset and relatively brief duration and by concurrent disturbances of attention, sleep-wake cycle and psychomotor behaviour.

The incidence of delirium in various Western studies in reported as 15–35% in aged 65 years and above, on admission to hospital.

Clinical Picture

The clinical features are usually similar to those in younger subjects, although elderly often fail to show the more elaborate hallucinations, oneiric thinking and confabulations.

Management

The identification and treatment of the underlying organic cause is **most important**. Supportive measures include maintenance of fluid and electrolyte balance, a balanced sensory environment and good nursing care.

(b) Dementia

It is the most *common irreversible psychiatric disorder* of the elderly. In Western countries its prevalence among people above 65 years range from 1.2 to 21.9% depending upon the sample studies and also the diagnostic criteria used.

The most striking features involve deterioration of memory, general intellectual and specific cognitive capacities and social functioning. Consciousness remains full and clear.

The *most common form of primary dementia* is Alzheimer's disease, which may have an onset in either the presenile or senile period (dividing line is age 65).

Table 1: Common Causes of Delirium in the Elderly

Drugs	Antihistaminics, neuroleptics, antiparkinsonian agents, hypnotics, diuretics, analgesics, digoxin.
Metabolic	Electrolyte imbalance, respiratory, hepatic or renal failure, endocrine disease (e.g. diabetes mellitus), vitamin B deficiency, hypothermia.
Infections	Urinary tract and chest infections and septicemia.
Cardiovascular	Cardiac failure, myocardial infarction, hypertension, hypotension, anemias, arrythmias, aortic stenosis.
Intracerebral	Cerebrovascular accidents (e.g. transient ischaemic attacks, cerebral thrombosis, embolism, haemorrhage, hypertensive encephalopathy, etc.), space occupying lesions, postconcussional states, postictal, encephalitis, meningitis.
Trauma	Fractures (especially femur, hip), surgery.

Restriction of activities

(i) *In early stage*: The primary dementias are invariable disabling, since the disorder involves first the higher order tasks such as work, handling finances, finding the way in public places, shopping or doing household chores.

(ii) *Later stage*: At a later stage, the simple self care tasks such as bathing, dressing, use of toilet, mobility, continence and feeding are affected. Life expectancy is considerably shortened because of infections.

The younger is the age of onset, more rapid is the deterioration, more the symptoms of parietal and temporal lobe dysfunctions (language disorganisation, aphasia, apraxia) and gait disturbances.

The correct diagnosis depends on the reliable history, investigations and special tests and mental status examination.

Differential Diagnosis

(i) Delirium

(ii) Depression

(iii) Others e.g., amnestic syndromes, hysterical amnesia, Ganser's syndrome etc.

Management

The treatment of cognitive impairment is first and foremost identification of reversible conditions, which are found in about 10–20% of investigated cases.

(a) ***Antipsychotic drugs***: The main indications are schizophrenia, paranoid states, mania, delirium and the restlessness, aggression or agitation associated with dementia or depression.

The following points should be kept in mind:

- Elderly are more sensitive to extrapyramidal side effects and tardive dyskinesia.
- Avoid use of drugs with more anticholinergic sideeffects e.g., thioridazine (risk of cardiotoxicity, urinary retention and delirium).
- Avoid use of high potency drugs (e.g., haloperidol) as they may cause parkinsonian sideeffects and akathisia.
- Use of selective D2 receptor antagonists e.g., Sulpiride, Pimozide or newer atypical agents such as risperidone, olanzapine etc. in low dosages are preferable.

(b) ***Antidepressants***: The following precautions should be kept in mind:

- Elderly are more prone to anticholinergic side effects, sedation, postural hypotension etc.
- Use antidepressants especially tricyclics in lower dosage range (because their half life is increased) and avoid tricyclics in acute recovery period following myocardial infarction or congestive heart failure, uncontrolled angina, arrythmias, pretreatment postural hypotension, narrow angle glaucoma and prostatic hypertrophy.
- Use safer drugs e.g., trazodone, mianserin, sertraline, fluoxetine, fluvoxamine, bupropion etc.

(c) ***Lithium*** is used with caution and the serum levels are maintained at the lower side (e.g. 0.6–0.7 meq/L)

(d) ***Anxiolytics and hypnotics***

- Use about half the doses of benzodiazepines as those in adults.
- Use drugs with a short plasma half life.
- Regular or prolonged use is avoided.
- Review the patients (on hypnotics) regularly

— the intermediate to short-acting benzodiazepines such Oxazepam, temazepam are the drugs of choice.

— use hypnotics with short half life (without hangover, dependence potential or severe withdrawal syndrome) e.g. zopiclone, zolpidem.

(e) ***Cerebral vasodilators*:** Dehydroergotoxine mesylate (Hydergine) is the most commonly used drug but other drugs used are papaverine, cyclandelate and isoxuprine.

(f) ***Cerebral stimulants*:** This class includes penthylenetrazol, pipradrol, methylphenidate and dexamphetamine. These are ineffective.

(g) ***Cholinergic drugs*:** The precursors choline and phosphatidyl choline (lecithin) are ineffective irrespective of dose or length of treatment. The cholinesterase inhibitor physostigmine infused intravenously may improve memory in ATD patients, as does the muscarinic against arecholine. Oral physostigmine produces a limited improvement. Rivastignine has been recently introduced but is costly.

(h) ***Electroconvulsive therapy*:** The following precautions should be taken:

— use unilateral than bilateral ECT

— use of a brief pulse stimulus preferred to reduce cognitive impairment

— use modified type ECT (with good premedication)

— use minimum number of ECTs

— use limited to depressed patients with psychotic symptoms.

(i) ***Psychotherapy*:** It is of value in patients with lowered self-esteem, loss of customary role.

II. FUNCTIONAL PSYCHOSES

(a) Disorders Resembling or Associated with Schizophrenia (paranoid syndromes)

History

Roth (1955) used the term ***late paraphrenia*** while Fish (1960) preferred the term ***senile schizophrenia.***

Epidemiology

— Prevalence is about 0.2 to 0.3% of population over 65. More common in females.

— 4% of schizophrenic disorders in men and 14% in women arise after age 65.

— 5.6% of all psychiatric first admissions after age 65 are for paranoid psychosis.

Etiology

(*i*) *Genetics*: There is increased risk of schizophrenia in relatives of late paraphrenics.

(*ii*) *Organic causes*: Cerebral lesions especially of temporal lobe and diencephalon.

(*iii*) *Sensory deficits*: About 30–40% of paranoid psychotics have impaired hearing.

(*iv*) *Personality*: The patients are often withdrawn, suspicious, sensitive premorbid personality—paranoid or schizoid type.

(*v*) *Environmental*: The precipitants often more uncover the pre-existing psychosis.

Clinical Picture

— usually insidious onset as a well-organised paranoid delusional system.

— hallucinations may not be present or may be bizarre.

— mood is often congruous.

— personality is frequently well-preserved.

Differential Diagnosis

(*i*) Organic cerebral disease

(*ii*) Depression

Management

Neuroleptics and depot preparations may be indicated. Electroconvulsive therapy may be used.

(b) Mood (Affective) Disorders

(i) *Depressive Disorder*

Epidemiology: In India, depression is believed to be the ***commonest psychiatric illness*** among psychogeriatric population in contrast to Western countries where dementia is the commonest diagnosis.

About 10% of the over 65s in a community has signs of depression whereas older people living in residential care, 30–50% may be depressed.

The prevalence of depressive episodes is increased in females or if there is past history of depressive or neurotic disorder, personality deviation, social isolation, presence of physical ill health or there was a early loss of parent.

Etiology

(*i*) *Genetic factors*: There is much less evidence of familial incidence in late onset (over 50) compared with early onset.

(*ii*) *Organic factors*

— Cerebrovascular disease may act as a precipitant of depression.

— The causes of symptomatic depression may include antihypertensive drugs, benzodiazepines, neuroleptics, hypothyroidism or potassium deficiency.

(*iii*) *Personality*: Neurotic depression may be related to obsessional premorbid personality.

(*iv*) *Environmental factors*

— A significant excess of loss in late onset depression compared with early onset.

— In the year following death of spouse, there is increased incidence of suicide.

Clinical Picture

Agitation is much common than retardation. There may also be delusions of guilt, poverty, nihilism and persecution. There is a tendency to answer "don't know" rather than confabulate (Pseudodementia). Suicide is common in isolated men, danger in elderly depressed, socially isolated men.

Differential Diagnosis

The common disorders from which depression needs differentiation are discussed in chapter 11 on "Mood Disorders".

(ii) ***Bipolar Disorder-Mania*:** About 5% of affective episodes in over 65, are diagnosed as mania or hypomania. Mixed affective states are common.

Hypomania in elderly is characterised by

- irritability
- garrulous, anecdotal speech with little flight of ideas.
- paranoid or sexual delusions or preoccupations.
- patients claim to be very happy but appear tense, irritable and miserable without any infectious gaiety—"*Miserable Mania.*"
- it may present as confusion or possibly delirium.
- mixed mood states quite commonly occur with elements of irritability, pessimism, expansiveness and self-denigration.

Management

(a) ***General Measures*:** Hospitalization; full assessment of social factors, isolation, housing and family support; investigate and treat any intercurrent physical illness.

Table 2: Specific Problems in Elderly and their Management

Problem	*Management*
Forgets Medication	Neighbour or Care assistant sets out medication, Calendar box.
Forgets Familiar People	Explain to the people; show them how to introduce themselves naturally
Forgets, Repeats Himself	Help carer to introduce distraction, give carer relief.
Aggression	Try and work out causes; if driven by paranoid ideas, treat with medication; If not, understand antecedents if possible, and counsel carer accordingly.
Shouting	Try to find the cause (e.g. deafness, pain); if no cause found, try to reward silence with caring attention, and refrain from rewarding shouting. But apparently causeless shouting is one of the most difficult problems in dementia.

Contd...

Problem	*Management*
Night time Restlessness	Reduce daytime boredom, avoid too early bedtime; maintain clear diurnal rhythm in household; provide commode for nocturnal micturition, carefully medication (Not Benzodiazepine).
Incontinence	Reduce obstacles to continence (difficulty in getting out of chair or walking, awkward geography of house, complicated clothing, constipation), regular reminders or actual taking to toilet; Pads often confuse a potentially continent; avoid them if possible.
Wandering Outside the House	Accept the risk; alternative daytime activity; change the door catches.
Disinhibition of Sexual Behaviour	Encourage open discussion of problem among carers, so that they can give clear unembarrassed cues to patient at the time to orient him to the social context.
Emotional Reactions to Disability, Clinging, Anger, Stubborn Catastrophic Reaction	Reduce stresses of patient. Introduce change very gradually, preferably through one trusted person. Introduction to supportive friendly environment is adherence to Familiar Routines, often very helpful.
Financial Incompetence	Arrange to draw pension (as appointee).
Problems in the carer (too frail, demented, depressed, resents)	Major changes in the situation (through help into the home) or move the patient out of home.

(b) ***Specific Treatment***

— *Medication.* Tricyclic antidepressants. The use of nonsedating antidepressants and with low anticholenergic side effects e.g., Sertraline, bupropion, venlafaxine, desipramine and fluoxetine may be preferred. Tranquilizers are indicated in agitation, delusional depression or manic episodes.

— *Electroconvulsive therapy (ECT).* less hazardous than drugs.

— *Social measures.* Rehabilitation measures are vital in all cases. Occupational therapy, home assessment, improvement of social support and development of "second careers" are all of great importance.

Outcome. *Poor* prognostic indications are

— Late Onset (onset after age 70)

— Organic brain disease

— Serious Physical disease

— Senile habits

— Uninterrupted depression for more than 2 years.

III. NEUROSES

(a) Anxiety States

Epidemiology

Prevalence of anxiety disorders in the general population is around 10% of elderly women (65 years or more) with the majority being phobias especially agoraphobias.

Clinical Picture

— *Emotional component*: fear, tension, dread, irritability and worried apprehension.

— *Behavioural component*: distractibility, complains, reassurance seeking.

— *Somatic component*: Feelings of respiratory restriction, palpitations, tremors, dizziness, headache, chest pains, rapid breathing, vomiting, coughing, rapid pulse and sweating.

Diagnosis

The common physical conditions which are likely to be misdiagnosed are

— hyperthyroidism

— hypoglycemia

— excessive intake of caffeine

— silent myocardial infarction

— small stroke or cerebral ischaemic attack

— withdrawal symptoms of sedatives, hypnotics or alcohol or antihypertensives

— other drugs e.g. bronchodilators, L-dopa.

The other conditions which may need to be ruled out are dementia, depression, phobic states etc.

Management

(*i*) ***Reassurance and explanations*** to the patient, supportive psychotherapy; strengthening of the involvement in social network and reduction of environmental threats.

(*ii*) ***Medication***

— tricyclic antidepressants and MAO inhibitors in phobic anxiety.

— benzodiazepines in generalised anxiety disorder especially those with shorter (4–8 hours) half life e.g. zolpidem, zopiclone, oxazepam, lorazepam.

— barbiturates are avoided.

— propranolol reduces somatic anxiety but add to sleep disturbance and is contraindicated in those with diabetes, cardiac failure or bronchial asthma.

(*iii*) ***Other measures*** include adequate exercise, curtailment of excessive day sleeping; regular bed time rituals; simple relaxation routines and formal behavioural strategies may also help.

GENERAL GUIDELINES FOR MANAGEMENT

The general principles of management include:

(*i*) Early, correct and full diagnosis of medical and social aspects.

* Assess at home, take a full history from patient, relatives, friends, co-workers etc.
* Assess the problem where it presents, assess local resources also.

(*ii*) Keep patient at home as long as possible.

* This reduces confusion and danger of institutionalisation and encourages utilisation of local resources.
* Family support is the most important factor here and families must themselves be supported, problems explained and discussed.
* Maximise home support with community nurses, social workers, practical help (meals on wheels, home helps etc.), ensure correct accommodation (warden controlled flat etc.).
* Outpatient clinic support of patient and relatives may be very helpful.
* Day hospital, day centre, luncheon club or Geriatric homes also aid in management.
* Admission to a short-stay psychogeriatric unit for full physical, psychological and mental assessment may be indicated. Multidisciplinary team work with clear definition of individual responsibilities may help.

Treatment

The possible modes of treatment are:

(a) Drugs

The following precautions should be taken

* Beware of overmedication or undertreatment.
* Assess physical condition (heart, lungs, kidneys, liver) and presence of other drugs (alcohol, antihypertensives etc.).
* Treat with the lowest effective dose.
* Use a limited range of familiar drug.
* Introduce medication slowly and carefully to avoid side effects and to increase compliance. Assess with plasma levels if available.
* If at home, give small quantities with each prescription and supply large written instructions.
* Explain treatment to relatives and involve relatives as appropriate.

(b) Psychotherapy

Patients may need long-term therapy but shorter individual sessions. It should pay attention to self-esteem and practical issues.

(c) Hospitalisation (Special geriatric units)

* Avoid institutionalisation
* Build and maintain interest on unit
* High staff-patient ratio
* Treatment of patients with respect

Emergency Presentations

* Physical illness.

* Effects of mediation.
* Breakdown of support system.
* Major change of environment.
* Paranoid illness.
* Attempted suicide.
* Agitation or panic attack.

Non-emergency Presentations

* Depression (Mild to moderate).
* Dementia (Mild to moderate).
* Multiple handicaps gradually intensifying.
* Neurotic illnesses.
* Alcoholism.

NURSING CARE

The elderly in nursing homes, geriatric centres, or community facilities require some special services and activities to live in dignity and comfort. These include:

- Maintenance of security and dignity and reinforcement of independence to the fullest extent appropriate to the patient's condition.
- Maintenance of the patient's orientation: name of the place, clock, calendar, and daily schedule in a convenient location.
- Encouragement of friendship and communication with staff and other patients.
- Arrangements for patients with visual defects to have assistance for safety.
- Staff members should speak clearly, slowly, and firmly to patients with hearing loss.
- If the patient suffers a painful condition, be especially sensitive to the need for medication to relative pain.
- Special attention to depression that may reach a stage of suicide risk when a personal or health crisis occurs.
- Observation of any specific interest in an activity reported to a staff member.
- Observation of eating habits, especially for food likes and dislikes, difficulty in chewing, or any fixed ideas about food. Report these to the dietitian.
- Report complaints about dentures so that patients can be seen by a dentist. Provide opportunity for women to visit a market.
- Care patients dress and grooming and praise on any accomplishment. Keep promises to patients faithfully.

17

Emergencies in Psychiatry

Definition

A psychiatric emergency is any disturbance in thoughts, feelings, or actions for which immediate therapeutic interventions are necessary.

General Principles

The principles underlying many psychiatric emergencies remain the same. The emergent task at hand becomes that of triage, evaluation, formulation, and disposition, keeping safety for the patient, others, and oneself always at the forefront. A careful, systematic approach can make good work practice routine, missing no detail. Always err on the side of caution and appropriate management of psychiatric emergencies saves lives.

A medical evaluation involves checking medications, vital signs, laboratory data and radiological evidence for medical causes of psychiatric morbidity. Once you have guaranteed safety and thoroughly ruled out medical causes for the presentation, consider psychiatric disorders.

A thorough mental status examination will guide diagnosis and sound care. Delirious patients require medical attention; suicidal patients demand protection; labile patients may need chemical or physical restraint; psychosis often impairs decision making capacity and ability to care for self; florid paranoia can endanger those in community.

We will briefly discuss some aspects of two psychiatric emergencies.

1. The Suicidal patient
2. The Violent patient

1. The Suicidal Patient

Introduction

Suicide is the termination of one's life intentionally.

According to 1999 data from the Centre for Disease Control and Prevention, suicide kills more people than homicide. Suicide was the eleventh cause of death (homicide was fourteenth), and the third leading cause of death between ages 15 and 24 years.

In India, suicide is the second leading cause of death among college students. The suicide rate for the young (15–24 years) was 10.3 deaths per 100,000 persons. Women attempt suicide more frequently than men (3:1), although men commit suicide more frequently (4.1:1.0).

Suicide and suicidal patient represent significant public health problem. The vast majority of people with suicidal intent have a major psychiatric diagnosis. It has been estimated that 90% or more of them can be shown to have a major psychiatric illness. A significant proportion, estimated as high as 70%, saw a physician within 30 days prior to their death. Nearly 50% had seen a physician in preceding week. It was noted that only 18% of suicidal patients communicated their intent to helping professionals, while 69% communicated their intent to an average of three close relatives or associates, 73% within 12 months of their suicide.

Risk Factors

Suicidal behaviour is the end result of a complex interaction of psychiatric, social and familial factors. An excellent mnemonic for the major risk factor is "SAD PERSONS," devised by Petterson et al.

Sex: Women are more likely to attempt suicide; men are more likely to succeed.

Age: Age falls into a bimodal distribution, with teenagers and the elderly at highest risk.

Depression: Fifteen per cent of depressive patients die by suicide.

Previous attempt: Ten per cent of those who have previously attempted suicide die by suicide.

Ethanol abuse: Fifteen per cent of alcoholics commit suicide.

Rational thinking loss: Psychosis is a risk factor, and 10% of patients with chronic schizophrenia die by suicide.

Social supports are lacking.

Organised plan: A well-formulated suicide plan is a red flag.

No spouse: Being divorced, separated, or widowed is a risk factor; having responsibility for children is an important statistical protector against suicide.

Sickness: Chronic illness is a risk factor.

Warning signs

- Suicidal talk by patient
- Preoccupation with death and dying.
- Signs of depression
- Behavioural changes
- Giving away special possessions
- Difficulty with appetite and sleep
- Taking excessive risks
- Increased drug use
- Loss of interest in usual activities

Assessment

It is often challenging to talk to patients about suicidal thoughts. It is important to remember that you cannot plant thoughts in the patient's head. If the patient has a plan, encourage him or her to speak with you about it.

A full current mental status examination is always a vital part of the assessment. Area of focus includes appearance, behaviour, mood, level of psychomotor activity and thought processes.

Factors to be considered in the assessment of the acutely suicidal patients include:

Assessment factors

Mental status, Suicidal ideation, Intent, Plans, Sadness, Hopelessness, Social withdrawal, Isolation, Anxiety, Agitation, Impulsivity, Insomnia, Psychosis, Prior high lethality attempts, Uncommunicative presentation, Recent major loss, Active substance abuse, Untreated mood, Psychotic or personality disorder, Common presentations include acute, chronic, contingent, and/or potentially manipulative suicidal patient. All are associated with anxiety for the care provider doing the assessment. Careful assessment, use of collateral information, and acceptance of predictive limitations can be helpful.

Para suicide

Self-destructive behaviour and nonfatal suicide attempts, although difficult to categorise, have been conceptualised as Para suicide.

Management

Suicide is a multi-dimensionally determined act, and the management of suicidal patient involves many treatment modalities.

Suicidal behaviour is a syndrome that cuts across rigid diagnostics lines.

The clinician must first confront his or her own feelings, however unpleasant or embarrassing, regarding the patient. Regardless of the underlying diagnosis, safety of the patient remains paramount. Principles of acute intervention begin with adequate supervision of the suicidal patient. If this supervision cannot be sufficiently accomplished in an outpatient setting, hospitalisation should be considered.

Should the patient be an active risk and unable to contract for safety, initiate one-to-one watch. Always look for methods the patient might use in the immediate vicinity to injure himself or herself. It is necessary to limit the patient's access to potentially self destructive methods like antidepressants medication, knives, firearms etc.

Treatment starts during the interview. A supportive style, with emphasis on encouraging the patient to share concerns, often proves therapeutic. Ensure medical clearance to expedite the initiation of pharmacotherapy. Routine tests include complete blood count with differential, electrolytes, liver function tests, thyroid function, and ECG and urine toxicology.

Management depends to a large degree on diagnosis. The aim of somatic treatment of suicidal patients to treat diagnosed psychiatric condition.

High risk diagnosis includes major depression, bipolar disorder, schizophrenia, alcoholism and substance abuse and borderline personality disorder. Co-morbid alcoholism increases risk in every diagnostic category.

Antidepressants, lithium, mood stabilisers, and ECT, as well as adjunctive benzodiazepines and antipsychotics, can significantly ameliorate underlying psychiatric diagnosis associated with suicide.

2. The Violent Patient

Violence by patients towards clinicians is not uncommon and is serious problem.

A number of studies indicate that approximately 10% of the patients seen in psychiatric hospitals manifested violent behaviour toward others just before being admitted to these hospitals, a finding that is true for private as well as public hospitals. Learning how to evaluate and manage violent patients is important not only for the safety of society and of patients in treatment settings but also the safety of mental health professionals.

Agitation is a state of increased mental excitement and motor activity. It may occur in a wide range of mental disorders. Common examples include hyperactivity, verbal abuse, threatening gestures and language. Unmanaged acute agitation can lead to violence. As such acute agitation is a psychiatric emergency because agitation often precedes violence and requires rapid intervention.

Literature review

A literature review on the topic of seclusion and restraint reached on the following conclusions concerning the risks and benefits of this intervention:

1. Seclusion and restraint are basically efficious in preventing injury and reducing agitation.
2. It is nearly impossible to operate a program for severely symptomatic individuals without some form of seclusion or physical or mechanical restraint.
3. Restraint and seclusion may have deleterious, physical and psychological effects on patients and staff.
4. Local non-clinical factors such as cultural biases, staff role perceptions, and the attitude of hospital administration, have a substantial influence on rates of restraint and seclusion.
5. Training in the prediction and prevention of violence, in self-defense, and in the implementation of restraint and/or seclusion is valuable in, reducing the rates and untoward effects of these procedures.

Definitions

Seclusion: The therapeutic isolation of a patient. It is of two types.

Open seclusion: Methods of open seclusion include quiet time alone in a patient's room, in an unlocked time-out room, or in a partitioned area. It represents the least restrictive form of seclusion.

Locked seclusion: The therapeutic isolation of a patient in a locked room designed specifically for the purpose of confining an agitated patient.

Restraint: A confining apparatus commonly composed of leather or canvas. When properly applied, restraints maximally restrict physical movement without threatening the body parts.

Indications for Seclusion or Restraint

- To prevent harm to others
- To prevent harm to the patient
- To prevent serious disruption of treatment environment
- As ongoing behavioural treatment

Contraindications specific to seclusion:

- The acutely suicidal patient (without constant observation)
- The patient with unstable medical status
- The delirious patient
- The self mutilating patient (without constant observation)
- The patient with a seizure disorder
- The developmentally disabled patient.

Guidelines for seclusion and restraint

- Not to be used to punish a patient or solely for the convenience of staff or other patients.
- Psychiatrist/staff must take into consideration the medical and psychiatric status of the patient.
- Psychiatrist/staff must follow written guidelines of institution.
- Adequate staff must be present (four) for implementation.
- When decision is made to use seclusion or restraint. Patient should be given seconds to comply by walking to the seclusion room.
- If patient does not comply, each staff member should grab a limb and bring patient backward to the ground.
- Restraint devices are applied, or patient is carried to the seclusion room by four staff member.
- Patient is searched for belts, pins, watches and other dangerous objects.
- Physician sees patient within 1 hour.
- Nursing staff should observe patient at least every 15 minutes.
- Meals, fluids and toileting should be provided at appropriate times and with caution.
- Patient should be gradually released from or restraint.
- Each decision, observation, and measurement, as well as care must be documented in detail in the patient's record.

Diagnosis

Rule out an organic mental syndrome, such as delirium or dementia. Obtain the patient's vital signs if possible. Abnormal vital signs suggesting autonomic abnormalities are the first clues of an organic disorder such as drugs or alcohol intoxication or withdrawal.

Is the patient paranoid and psychotic, with impaired reality testing? If the patient is psychotic and agitated, medication may be indicated immediately. Is there a treatable medical cause? Many medical conditions (for example, sympathomimetics, anticholinergics and digitalis) can precipitate episodes of agitation. Does the patient suffer from personality disorder that may make the patient prone to impulsivity or to excessive anxiety in response to stress?

Make a definite diagnosis so that a treatment plan can be developed.

Interviewing guidelines

If the conversation is possible, try to quiet the patient. It is important not to express overt anger or hostility. Do not be punitive. It is also important to remain non-confrontational and to let the patient know that you will listen empathetically to angry complaints and concerns and that you will be honest with the patient about limits and treatment. Be reassuring to patient and say that every one there is trying to help. Stay calm and straightforward as far as possible. If talking is not effective isolate the patient and avoid excessive stimulation. If patient requires medication for sedation use it appropriately.

Assessment

The time available for patient assessment will be dependent on the acuity of the presentation. For someone who is acutely agitated and an immediately danger to self or others, emergency measures must be taken to avoid harm.

Key points in assessment of acute agitation? Violence is outlined below:

1. Somatic conditions
2. Previous history of violence
3. Access to weapons
4. Current ideation, including content of delusions
5. Active substance abuse, including alcohol
6. Co-morbid antisocial personality disorder
7. Verbal threats
8. Premonitory physical signs (clenched fist, pacing)

Management

Behavioural, psychological and pharmacological interventions are used simultaneously. Protect yourself and the staff. Have a sufficient number of staff members present to restrain the patient if necessary. Physical restraint should be used if medications are ineffective and if violence or fight is impending. For violence, tranquilization may be necessary. Usually, benzodiazepines or antipsychotics are used. The drug choice is based on the available routes of administration, potency and the side effect profile.

If the patient is taking a specific drug or has a history of responding to a specific drug, use that drug again.

Table 1: Pharmacological Options in Emergencies for Violence

Drug	*Dose (mg)*	*Half-life (h)*
Lorazepam	2–4 mg im or orally; repeat every hour if im or every 4–6 hr if oral	10–20
Haloperidol	2–5 mg im every 1–4 hr with maximum daily dose of 20 mg	12–36
Olanzapine	10 mg im and repeat if necessary	34–38
Chlorpromazine	25 mg im every 4 hours with increased dose over 1–3 days	–

* im—intramuscularly.

A combination of the two drugs is safe and is often used.

In a survey, the most frequently used medications were haloperidol and lorazepam, often together and intramuscularly.

For long-term medication there is no one drug for treatment of violence because the underlying etiology for violence differs among patients. Early intervention is important to avoid further escalation to violence.

The new intramuscular formulations of atypical antipsychotics hold promise to quickly and efficaciously control acute agitation, without side effect burdens of older typical antipsychotics.

18

Cultural Bound Syndromes in India

Culture plays a decisive role in colouring the psychopathology of various psychiatric disorders. However, some psychiatric syndromes are limited to certain specific cultures. These disorders are called culture specific or culture bound syndrome. The last two decades have witnessed an increased interest in the cross cultural study of psychiatric disorders.

Culture specific syndrome or **Culture bound syndrome** is a combination of psychiatric and somatic symptoms that are considered to be a recognisable disease only within a specific society or culture. There is no objective biochemical or structural alterations of body organs or functions, and the disease is not recognised in other cultures.

The term culture bound syndrome was included in the fourth version of the *Diagnostic and Statistical Manual of Mental Disorders* (American Psychiatric Association, 1994) which also includes a list of the most common culture bound conditions. According to DSM-IV culture bound syndrome denotes **recurrent, locality-specific patterns of aberrant behaviour** and troubling experience that may or may not be linked to a particular DSM-IV diagnostic category. Many of these patterns are indigenously considered to be 'illnesses', or at least afflictions and most have local names. Although presentations conforming to the major DSM-IV categories can be found throughout the world, the particular symptoms, course, and social response are very often influenced by local cultural factors. In contrast, *culture bound syndromes* (CBS) are generally **limited to specific societies or culture areas and are localised, folk, diagnostic categories** that frame coherent meanings for certain repetitive, patterned, and troubling sets of experiences and observations (American Psychiatric Association, 1994 : 844)

Though no clear cut diagnostic criteria have been devised as of now, majority of CBS share the following characteristics:

- Categorised as a disease in that culture
- Widespread familiarity in that culture
- Unknown in other cultures

- No objectively demonstrable biochemical or organ abnormality
- Treated by folk medicine/traditional healers

In India, common culture bound syndromes are Dhat syndrome, Possession syndrome, Koro, Gilhari syndrome, Bhanmati, Compulsive spitting, culture bound suicide (sati, santhra), Ascetic syndrome, Jhinjhinia etc.

The chapter will discuss the Socio-demographic, clinical profile and nosological status of various culture bound syndromes in the Indian subcontinent.

1. Dhat Syndrome

Dhat syndrome is a clinical entity recognised both by general public as well as medical practitioners in which nocturnal emissions lead to severe anxiety and hypochondriasis, often associated with sexual impotence. Patient usually presents with various somatic, psychological and sexual symptoms. Patient attributes it to the passing of whitish discharge, believed to be semen (Dhat), in urine.

Dhat is derived from sanskrit word *'Dhatu'* meaning precious fluid. Susruta Samhita (ancient Indian text of surgery) has described 7 Dhatus in the body. Dhatus are elixir of the body. Disturbances of it can cause physical and mental weakness. Of all seven, Semen is considered to be the most precious. Charak Samhita (ancient text of Indian Medicine) describes a disorder resembling Dhat Syndrome by the name 'Shukrameha.' Shukra is the word used for sperms in Sanskrit. Another term denoting semen is 'Veerya' which in Sanskrit means bravery, valor and strength. This gives rise to belief that loss of excessive semen in any form e.g. masturbation, nocturnal emissions etc. is harmful. On the other hand, its preservation will lead to health and longevity. Thus the belief in precious and life-preserving properties of semen is deeply ingrained in Indian culture. The belief is further reinforced by traditional healers and perpetuated by friends and elders who had suffered from this syndrome.

- The term was first used by N.N. Wig in 1960.
- A whitish discharge is blamed by patient to be responsible for the physical and mental symptoms which patient suffer from. However there is no objective evidence of such a discharge. Sometimes patient also reports of foul smelling semen and less viscous semen. Patient complaints of following symptoms:
- Generalised weakness, aches and pains all over body.
- Tingling and numbness in various parts of body especially peripheries.
- Easy fatigue.
- Lassitude.
- Loss of appetite, weight loss, loss of attention and concentration.
- Excessive worrying.
- Panic attacks.
- Sadness of mood.

- Forgetfulness.
- Feelings of guilt (especially towards masturbation during adolescence).
- Sexual complaints are that of premature ejaculation and erectile dysfunction.
- In majority of cases there is absence of any physical illness like diabetes, local genital abnormalities, sexually transmitted diseases.

The syndrome is seen usually in people from lower socioeconomic strata who seek help from traditional healers before reaching Hospitals. Present all over the country. Also seen in surrounding countries like Sri Lanka (Sukra Prameha), Pakistan and even in China (Sen-k'uri). Concomitant psychiatric morbidity like depression, somatoform disorder, anxiety disorder may be present.

Treatment mainly consists of dispelling of myths by psychoeducation, reassuring the patient, treating any underlying psychiatric disorder, even symptomatic relief (of severe anxiety that these patients suffer) with the help of medications in initial stages of treatment is required to gain confidence of the patient.

2. Possession Syndrome

Diagnosable under dissociative disorders in ICD-10. Patient is possessed, usually by 'spirit/ soul' of deceased relative or a local deity. Patient speaks in a changed tone, even gender changes at times if the possessing soul is of opposite sex. This is usually seen in rural areas or in migrants from rural areas.

Majority of these patients are females who otherwise don't have any outlet to express their emotions. Treatment includes careful exploration of underlying stress which precipitated the possession attack. Also to decrease any secondary gains patient may be getting from this behaviour.

Treatment antidepressants and anxiolytics are helpful in certain cases. Syndrome is seen in all parts of India. Many religious shrines hold special annual festivals where hundreds of people get possessed simultaneously. These people are looked upon as special by their families and villages which reinforce the secondary gains.

3. Culture Bound Suicide

Various types of culture bound (culturally sanctioned) suicides are as follows:

(a) *Sati*: self-immolation by a widow on her husband's pyre.

According to Hindu mythology, Sati the wife of Dakhsha was so overcome at the demise of her husband that she immolated herself on his funeral pyre and burnt herself to ashes. Since then her name 'Sati' has come to be symptomatic of self-immolation by a widow.

Was seen mostly in Upper Castes notably Brahmins and Kshatriyas.

Banned in India since 19th century.

Only one known case since 1904 (in Rajasthan).

(b) *Jouhar*: Suicide committed by a women even before the death of her husband when faced by prospect of dishonour from another man (usually a conquering king)

- Most notable example is Rani Padmini of Chittor (Rajasthan) to evade the invading army of Sultan from Delhi in 15th century
- More recently, hundreds of women killed themselves by jumping in wells during (1947) partition of India to avoid rioters

(c) *Santhara/Sallekhana*: Voluntarily giving up life by fasting unto death over a period of time for religious reasons to attain God/Moksha

- Seen in Jain community who celebrates these events as religious festivals
- Person initially takes liquids, later even refusing to take them
- Recently 4 cases were reported from Rajasthan

4. Koro

- Seen in northeastern states like Assam
- Fear of genitalia retracting into abdomen leading ultimately to death
- Seen in both sexes
- Person applies external retractors to the genitalia in form of clamps, chains etc. to avoid it retracting back
- It may occur as epidemics
- Described as a syndrome in ICD-10 and DSM-IV

5. Ascetic Syndrome

- First described by Neki in 1972
- Appears in adolescents and young adults
- Characterised by social withdrawal, severe sexual abstinence, practice of religious austerities, lack of concern with physical appearance and considerable loss of weight

6. Jhin Jhinia

- Occurs in epidemic form in India
- Characterised by bizarre and seemingly involuntary contractions and spasms
- Nosological status unclear

7. Bhanmati Sorcery

This CBS is seen in South India. It is believed to be due to psychiatric illness i.e., conversion disorder, somatization disorder, anxiety disorder, dysthymia, schizophrenia etc.

- Nosological status unclear

8. Suudu

It is a culture specific syndrome of painful urination and pelvic "heat" familiar in South India, especially in the Tamil culture. It occurs in males and females. It is popularly attributed to an increase in the "inner heat" of the body often due to dehydration.

It is usually treated by the following:

1. Applying a few drops of sesame oil or castor oil in the navel and the pelvic region
2. Having an oil massage followed by a warm water bath
3. Intake of fenugreek seeds soaked overnight in water

   The problem has also been known to exist in other parts of South India and the methods of treatment are also similar

9. Gilhari Syndrome

- Characterised by patient complaining of small swelling on the body changing its position from time to time as if a gilhari (squirrel) is tavelling in the body
- Reported from Rajasthan (Bikaner)
- Not much literature available
- Nosological status not clear

10. Mass Hysteria

- Short lasting epidemics of Mass Hysteria where hundreds of thousands of people were seen to be believing and behaving in a manner in which ordinarily they won't.
- E.g. the God Ganesha's idols drinking milk all over India in 2006 which lasted for almost a week.

A report by Choudhary et al., in 1993—gives description of an atypical hysteria epidemic in a tribal village of the State of Tripura, India. Twelve persons, eight female and four male, were affected in a chain reaction within a span of ten days. The cardinal feature was an episodic trance state of 5 to 15 minutes duration with restlessness, attempts at self-injury, running away, inappropriate behaviour, inability to identify family members, refusal of food and intermittent mimicking of animal sounds. The illness was self-limiting and showed an individual course of one to three days duration.

19

Legal and Ethical Issues in Psychiatry

All of us are bound by the legal system of the land and the law of the country treats everyone to be equal. However, under certain situations where a citizen is incapable of carrying out the responsibilities of a normal citizen by reason of unsoundness of the mind, the law takes a different stand. A person with unsound mind has different provisions under some particular situations. Why should there be different provisions for the mentally ill:

1. Some mentally ill persons lack insight into their illness and under the influence of the illness they are incapable to evaluate the consequences of their actions.
2. Society's attitude towards the mentally ill has not remained favourable, they are generally ignored and looked down upon.
3. Some mentally ill, by reason of their mental illness develop suicidal or homicidal tendencies; these tendencies may be different from the ones experienced by the normal people. Law of the land takes a different stand for such unlawful activities.
4. Mentally ill persons are vulnerable to exploitation. Others often take advantage of their lack of understanding or subnormal intellect.
5. Because of subnormal intellect or otherwise being incapable to manage their affairs by reason of mental illness, they require someone else to look after them or manage their affairs.

Due to these reasons it becomes necessary to have separate provisions for the mentally ill. The situations under which law interacts with psychiatry make the domain of legal or forensic psychiatry. Forensic psychiatry deals with the legal aspects of psychiatry. The law and the practice of psychiatry is very wide and has two dimensions. The first aspect covers the practice of psychiatry as defined through the constitutional, statutory and regulatory laws. These include issues related to the rights of the patient while giving consent to treatment, doctor's responsibility, voluntary and involuntary treatment, professional negligence, confidentiality, record keeping, etc. The second aspect relates to psychiatry serving the needs of the legal system. These include issues concerning with patient's competency, to stand trial, his criminal responsibility, civil responsibility like testamentary capacity, marriage, competency to manage one's own property, etc. The law comes in contact with psychiatry at many points. These can be broadly classified into two groups:

1. Admission, detention and care of the mentally ill in a Psychiatric Hospital or Psychiatric Nursing Home.
2. General laws of the land in relation to the mentally ill, such as, criminal responsibility, civil responsibility of a mentally ill.

Laws Related to the Care of Mentally Ill

Historically, mentally ill were variously treated, neglected, tortured and looked down upon. They were considered to be the victims of the curse... in league with evil spirits... possessed by ghosts and so on. 16th century onwards, there appeared a trend of isolating them in order to protect rest of the society. Lunatic asylums for their custodial care which soon degenerated into the institutions to inflict all kinds of cruelty on the inmates were built. With the dawn of 20th century, democratisation of the society, increased scientific knowledge, human rights movement and many other factors contributed to improve the status of the mentally ill. More elaborate laws related to their custodial care (known as Lunacy Acts/Mental Health Act) came and special provisions in civil and criminal laws related to mentally ill enacted in accordance with their status. These laws are as follows:

Mental Health Act

The Indian Mental Health Act, 1987 replaces the Indian Lunacy Act of 1912. The Mental Health Act provides not only for the custodial care of the mentally ill, it also provides guidelines for the protection of their human rights, safeguards against any kind of exploitation and proper monitoring of their care while they are inmates in the psychiatric hospitals. This Act abandons the older and derogatory terms used in Indian Lunacy Act of 1912 such as Asylums and Lunatics and are replaced by the terms psychiatric hospitals and mentally ill persons. New provision for admission to the hospital has been added wherein an uncooperative or unable to give consent patient can be admitted under special circumstances. Patient can be admitted on voluntary basis and under reception order as well. The Act also provides for creation of Mental Health Authority both at central and state level to monitor and regulate mental health activities. The Act extends to whole of India unlike the Indian Lunacy Act, 1912, which did not apply to Jammu and Kashmir state.

Criminal Responsibility

A crime committed by a mentally ill may be the result of delusional experiences the patent was having at the time of committing the crime. Can such a patient be held responsible for his action when he is living entirely in a different world dominated by his delusions. A British example illustrates it, which is popularly known as McNaughten's Rule(s). McNaughten was a young Scotsman who harboured delusions against the then British Prime Minister Sir Robert

Peel that the Prime Minister wanted him to be killed. He followed the Prime Minister wherever he went and decided to kill the Prime Minister before he could harm him. In an attempt to kill the Prime Minister, McNaughten shot at Edward Drummond, Prime Minister's secretary by mistake.

The jury, after testification by 9 physicians, found McNaughten 'not guilty by reason of insanity.' Queen Victoria, Sir Robert Peel and other prominent persons were outraged. Following this, 15 prominent judges were invited by the House of Lords. They were asked to respond to a series of questions on criminal responsibility. The answers, which are immortalised in the history of Forensic Psychiatry, have now come to be known as *McNaughten's Rule(s).*

In slightly modified from McNaughten's Rule(s) are adopted by many countries even now. It has been embodied in the Indian Criminal law, in Section 84 of Indian Penal Code (Act 45 of 1860). This states:

'Nothing is an offence, which is done by a person, who at the time of doing it, by reason of unsoundness of mind, is incapable of knowing the nature of the act or that he is doing what is either wrong or contrary to law.'

So, a mentally ill person is not protected *ipso facto*. He must satisfy the above mentioned rule. As per law, an *idiot* (mentally retarded person) is not liable in criminal law. The law classifies 'criminal lunatics' into three classes:

1. An undertrial, who cannot stand trial because of mental illness.
2. A person who is guilty but insane.
3. Criminals who later become mentally ill.

A person who is guilty but insane is a criminal lunatic who is acquitted by reason of insanity but detained in a psychiatric hospital (lunatic asylum, mental hospital) for treatment.

Suicide and Suicidal Attempt

Suicide is a type of *deliberate self-harm* and is defined as a human act of self-intentioned and self-inflicted cessation (death). It ends with a fatal outcome.

Suicidal attempt is an unsuccessful suicidal act with a non-fatal outcome. It is believed that 2–10% of all persons who attempt suicide, eventually complete suicide in the next 10 years.

A *suicidal gesture*, on the other hand, is an attempted suicide where person performing the action never intends to die by the act. However, some of these persons may accidentally die during the act. Attempted suicide is more common in women while completed suicide is 2–4 times common in men.

Under Indian law, suicide and suicidal attempt are punishable offenses. Section 309 of Indian Penal Code (IPC) states that *'whosoever attempts to commit suicide and does any act towards the commission of such offense, shall be punishable with simple imprisonment for a term which may extend to one year and shall also liable to fine.'*

Section 309 of IPC was repealed by the Supreme Court of India in 1994. However, in

March 1996, a five judge constitution bench of the Supreme Court again made 'attempted suicide' a punishable offense.

Civil Responsibility

It is presumed that every person is sane and the contrary must be proved. This applies to both civil and criminal proceedings in the court of law. Civil responsibilities of a mentally ill person is generally questioned under the following situations:

1. Testamentary Capacity

A person can will his property to someone before his death and this testamentary disposition is regulated by the Indian Succession Act (Act 39 of 1925). The following points need to be observed at the time of testamentary disposition:

(*i*) The will must be in writing, it cannot be a just verbal communication though it is not necessary to register it.

(*ii*) The testator should sign the will in the presence of at least two witnesses.

(*iii*) A beneficiary from the will cannot be a witness.

(*iv*) The testator appoints an executor who will execute the will after the death of the testator.

(*v*) The testator is free to revoke the will any time before his death.

(*vi*) A will is effective only after the death of the testator.

(*vii*) Every person with sound mind, not being a minor, may dispose of his property by will (Section 59, Indian Succession Act, 1925).

(*viii*) No person can make a will while he is in such a state of mind, due to whatever reason (intoxication, illness or other cause) that he does not know what he is doing.

Sometimes a doctor is required to examine the state of mind of the testator. While examining such a person the doctor should keep the following points in mind:

(*i*) Whether the testator knows the act he is doing and its consequences.

(*ii*) Whether he is aware of the extent of his property.

(*iii*) Whether he knows his exact relationship with the beneficiaries and the likely consequences of the will.

(*iv*) If the testator is seriously ill he must be made to read out aloud the will in the presence of the doctor.

(*v*) The doctor has to ensure that the testator is not under the influence of any kind of intoxicant while making a will.

(*vi*) Testator should not be under the influence or pressure of some other person while making a will.

(*vii*) A mentally retarded person, who is incapable of understanding the nature of the act, cannot make a will.

(*viii*) Deaf, dumb and blind persons who understand the nature of the act can make a will.

(*ix*) A mentally ill person in lucid interval wherein he is capable of understanding the nature of the act, can make a will.

2. Adoption

Any Hindu male who is of sound mind and not a minor can adopt a child with the consent of his wife, provided she is of sound mind.

Any Hindu unmarried female with sound mind and not being a minor can adopt a child. If she is married, her husband who is of sound mind can adopt a child.

A person capable of giving in a child for adoption should be of sound mind.

This is as per Hindu Adoptions and Maintenance Act (Act 78 of 1956).

3. Witness

Under the Indian Evidence Act, 1872 a lunatic is not competent to give evidence if he is prevented by virtue of his lunacy from understanding the questions put to him and giving rationale answers to them (Section 118). However, such a person can give evidence during a lucid interval; on the discretion of the judge and the jury.

4. Marriage

According to the Hindu Marriage Act (Act 25 of 1955), Section 5(ii) introduced by Act 68 of 1976, the conditions for a Hindu marriage are that at the time of marriage, neither party,

(a) Is incapable of giving a valid consent due to unsoundness of mind.

(b) Though capable of giving consent, has been suffering from mental disorder of such a kind or of such an extent as to be unfit for marriage and procreation of children; or

(c) Has been subject to recurrent attacks of insanity or epilepsy.

A marriage solemnised in contravention with this condition shall be voidable and may be annulled by a decree of nullity under Section 12 of the Act. If a consent for marriage is obtained by fraud .... 'as to any material fact and or circumstance concerning the respondent,' for example, the fact of the mental illness or treatment for the same.

Divorce can be granted under Section 13 of the Act on petition presented by either spouse on the ground that the other party 'has incurably of unsound mind, or has been suffering continuously or intermittently from mental disorder of such kind and to such an extent that the petitioner can not reasonably be expected to live with the respondent.

Narcotic Drugs and Psychotropic Substances Act (NDPSA), 1985

This Act replaces the earlier Acts, the Opium Act, 1857; Dangerous Drugs Act, 1930 and the Opium and the Revenue Laws Act, 1950. NDPSA came into force on 16th September 1985 to regulate production, possession, transport, import, export, sale, purchase or use of any narcotic drug or psychotropic substance. The Act provides for punishment of Rs. one lakh of fine with rigorous imprisonment of 10 years.

The Persons with Disabilities Act, 1995 *(Equal Opportunities, Protection of Rights and Full Participation)*

This Act provides for inclusion of mental illnesses in the purview of disability. Section 2(i) recognises Mental Illness and Mental Retardation as a disability.

A Medical Board to issue a disability certificate comprises of (i) medical superintendent of the hospital or his nominee, (ii) a physician and (iii) a psychiatrist. The board so constituted is competent to issue disability certificate.

Psychiatric disability is measured with Indian Disability Evaluation and Assessment Scale (IDEAS). Benefits are available for disability above 40%. Four mental illnesses are included for disability; these are Schizophrenia, Bipolar Disorders, Obsessive Compulsive Disorders and Dementia.

Mental Retardation Disability is measured with Intelligence Quotient (IQ). These are: Mild (IQ 50–69) 50%, Moderate 75%, Severe 90% and Profound 100%.

This Act is also helpful in psychosocial rehabilitation.

Ethical Issues in Psychiatry

In practice of psychiatry ethics includes the safeguards of the psychiatric profession and ensure high standards of professional conduct. Patient welfare is of utmost importance wherein autonomy of patient has to be maintained. Exploitation of patient in any form is unethical. Treatment has to be humane with professional competence on the part of the treating psychiatrist. Nothing has to be hidden from the patient and no discrimination has to be observed because of caste, creed, religion or any other factor. Confidentiality of the patients' problems has to be maintained and his secrets are not to be divulged to others.

Medical Malpractice

Guiding principle is, "no person shall be harmed by the act of other person." In case of medical practice, no harm is to be caused to the patient by the treatment of the treating doctor. Medical malpractice is a non-criminal civil wrong arising from the patient or caregiver's perception that the highest standards of medical care were not maintained. This may be the alleged negligence on the part of the treating practitioner. There are four elements of malpractice:

- Duty was not properly performed by the treating physician.
- There was a deviation from standards of the treatment.
- As a consequence a damage occurred.
- And that this was the direct causation.

For Any Charge of Malpractice

- Doctor-patient relationship: putting the doctor in a caring role
- Acceptability and standard of care provided
- Deviation from standards and norms

- In psychiatric practice: invasion of privacy, involuntary admission, seclusion, restraint (human right violation of the patient) forcibly administering treatment.

To Avoid Possibility of Malpractice

There should be a good doctor-patient relationship. Psychoeducation of caregiver and the patient is essential and they should be informed of possible outcome of the treatment in clear terms. There should be a continuing dialogue with caregivers should be there. The treating doctor should adhere to standards of caring and should follow only the accepted norms. Proper documentation of the treatment should be there. It may be referred to later on.

20

Psychopharmacology

I. ANTIPSYCHOTIC DRUGS

Also called *Major tranquilizers* or *neuroleptics* or *antischizophrenic drugs* or *ataractics* (see Table 1).

Table 1: Selected Antipsychotic Drugs Dosages

Class/Generic name	*Trade name*	*Dose equivalent (mg)*	*Usually daily oral dose (mg)*	*Parenteral single dose (mg)*
I. Phenothiazines				
a. Aliphatic				
(*i*) Chlorpromazine hydrochloride	Chlorpromazine	100	200–600 (up to 2000 mg)	25–100 (I/M)
b. Piperazine				
(*i*) Trifluoperazine	Espazine Trinicalm Neocalm, Trazine	2.4–3.2	5–40 (I/M)	1–2
(*ii*) Fluphenazine decanoate	Anatensol Fludecon Prolinate	0.61	10 mg of oral fluphenazine = 12.5–25 mg/ 2 weeks of fluphenazine decanoate	25–50 (I/M every 2–4 weeks)
(*iii*) Flupenthixol	Spenzo, Fluanxol	20–40		(I/M every 2–4 weeks)
(*iv*) Clopenthixol	Clopixol	100–200		(I/M every 2–4 weeks)
c. Piperidine				
(*i*) Thioridazine hydrochloride	Mellaril Thioril Sycoril Thiozine, Melozine	90–104	200–600	—

Contd...

Table 1: Selected Antipsychotic Drugs Dosages (*Contd...*)

Class/Generic name	*Trade name*	*Dose equivalent (mg)*	*Usually daily oral dose (mg)*	*Parenteral single dose (mg)*
II. Butyrophenones				
(*i*) Haloperidol	Halopidol Senorm Trancodol Serenace	1.1–2.1	2–12	2–5 (I/M or I/V)
(*ii*) Haloperidol decanoate	Senorm L.A. Serenace L.A.		10 mg/day oral haloperidol =10–200 mg/4 weeks of decanoate	100–200 (I/M every 4 weeks)
III. Thioxanthenes	**Not available/in use**			
IV. Diphenylbutyl Piperidines				
(*i*) Pimozide	Orap, Mozep	—	2–10	—
V. Dibenzoxazepine				
Loxapine	Loxapac	10	20	12.5–50 (I/M)
VI. Indole Derivatives	**Not Available/in use**			
VII. Dibenzodiazepine				
Clozapine	Sizopin, Lozapin	—	200–900	—
Olanzapine	Oleanz, Oliza, Olanex Olan, Olpine, Meltolan	—	5–20	—
VIII. Substituted benzamides	**Not Available/ in use**			
IX. Rauwolfia alkaloids	**Not Available/ in use**			
X. Miscellaneous				
Risperidone	Sizodon, Risdone, Rispid, Aprizol,	—	2–10	—
Aripiprazole	Aria, Arive Aripra, Arip	—	10–30	—
Quetiapine	Quitipin, Qutan	—	100–800	—
Remoxipride		30–60	150–600	—
Ziprasidone	Zipsidon	—	80–160	—
Amisulpride	Amazeo, Soltus, Sulpitac	—	200–1200	—
Levosulpride	Levazeo, Nexipride	—	100–600	—

Mechanism of Action of Antipsychotics

Blockage of dopamine receptors in ***caudate nucleus*** and the ***limbic system.*** The blockage of dopamine receptors (**D2 receptors**) in the mesolimbic system, thus resulting in increased dopamine turn over rate, produce antipsychotic effect. The blockage of dopamine receptors, resulting in increased dopamine turn over rate in caudate nucleus products parkinsonism like syndrome, which can be countered by antiparkinsonian drugs.

Adverse Effects – See Table 2.

Table 2: Adverse Effects of Antipsychotic Drugs and their Management

Type	*Side effect*	*Mechanism of action*	*Management*
A. Extrapyramidal Side Effects	1. Acute Dystonia (commonly Opisothotonus or torticollis, Oculogyric crisis).	Dopaminergic receptor (D2) blockade in Striatal system.	Antiparkinsonian anticholinergics, benzodiazepines. (Sometimes change in medication, or lowering dose).
	2. Akathisia (verbal or motor restlessness).	Dopaminergic receptor (D2) blockage in Striatal system.	Benzodiazepines, beta-blockers or antiparkinsonian (Sometimes lowering dosage, stopping or changing medication).
	3. Parkinsonian symptoms (Pseudoparkinsonism, akinesia, rigidity, tremors).	Dopaminergic receptor (D2) blockage in Striatal system.	Benzodiazepines, beta-blockers or antiparkisonian (amantadine may also be used).
	4. Rabbit Syndrome (chewing type movement as of a rabbit)	Dopaminergic receptor (D2) blockage in Striatal system.	Benzodiazepines, beta-blockers or antiparkinsonian (Sometimes lowering dosage, stopping or changing medication)
	5. Tardive Dyskinesia (bucco linguo laryngo-masticatory dyskinesia) (Risk more in elderly, females, brain damage, increased dose and duration of therapy. use of antiparkinsonians)	Post Synaptic Dopamine (D2) receptor supersensitivity (Noradrenergic hyperactivity) Not reported with clozapine (an antipsychotic drugs without extrapyramidal side effects).	Prevention best. Medications e.g. cholinergics (physostigmine, lecithin, Choline), reserpine, levodopa, benzodiazepines, lithium etc. and lastly the neuroleptics.
	6. Neuroleptic Malignant Syndrome	Not known	Dantrolene (1 mg/kg up to 10 mg/kg/day), Bromocriptine, Levo dopa, anticholinergics, ECT.
B. Other Neurological Side Effects	1. Seizures	Decreased seizure threshold	Use drugs with no or minimal effects on seizure threshold (e.g. haloperidol, Pimozide, trifluoperazine). Rarely, add anticonvulsant (carbamazepine, phenytoin)
	2. Sedation	Blockage of alpha-adrenergic receptors.	Use butyrophenones or pimozide; Give single dose at night. (Gradually tolerance develops).

Contd...

Type	*Side effect*	*Mechanism of action*	*Management*
	3. Depression (Pseudodepression)	Blockage of catecholamine receptors (noradrenergic and serotoninergic) in brain. Low dose of antidepressants or ECT.	Rule out Pseudoparkinsonism (or add antiparkinsonian drugs).
	4. Hallucinosis	Not known (Sedation is an important factor)	Decrease dose of drugs or change to one with minimum sedation or decrease the dose of antiparkinsonian drugs.
	5. Increased salivation (with clozapine)	Not known	Use antiparkinsonian drugs or stop drug.
C. Autonomic Side Effects			
a. Anticholinergic	1. Dry mouth	Blockage of Muscarinic Cholinergic receptors	Rinsing of mouth with water (avoid candy as caries may result). Pilocarpine (2%)
	2. Constipation	Blockage of Muscarinic Cholinergic receptors	Laxatives: change in diet; usually tolerance develops.
	3. Urinary retention	Blockage of Muscarinic Cholinergic receptors	Rule out benign hypertrophy of prostate. Bethanecholine (25–50mg tid) or catheterization Tolerance develops Stop anticholinergic antiparkinsonian or change antiparkinsonian or change antipsychotic.
	4. Cycloplegia	Blockage of Muscarinic Cholinergic receptors	Usually none. Sometimes Pilocarpine (2%)
	5. Mydriasis	Blockage of Muscarinic Cholinergic receptors	Usually none. Sometimes Pilocarpine (2%)
	6. Anticholinergic delirium	Blockage of Muscarinic Cholinergic receptors	Physostigmine (1–2 mg I/M) Diazepam Use neuroleptics with minimal anticholinergic effects and stop anticholinergic antiparkinsonians.
	7. Cholinergic crises	Blockage of Muscarinic Cholinergic receptors	Atropine
	8. Tachycardia	Blockade of Muscarinic Cholinergic receptors	Start in low dose. Prefer neuroleptics such as haloperidol.
b. Adrenergic blockade	1. Orthostatic hypotension	Blockade of alpha adrenergic receptors	Usually none, change in posture gradual, raise bed

Contd...

Type	Side effect	Mechanism of action	Management
	2. Impaired ejaculation and impotence	Blockade of alpha adrenergic receptors.	Decrease dose or change drug.
c. Combined	Temperature dysregulation.	Both antimuscarinic and alpha adrenergic blockade.	Stop drug if hyperthermia, adequate fluids, avoid exertion.
D. Allergic			
a. Hepatic	Cholestatic jaundice	Hypersensitivity reaction	Stop drug, benign course, supportive care, change drug.
b. Dermatological	1. Maculopapular skin eruptions	Hypersensitivity reaction	Discontinue drug and add antihistaminic e.g. diphenhydramine. Start drug from another class of antipsychotics (e.g. haloperidol)
	2. Photodermatitis (more with chlorpromazine)	Not known	Avoid sunlight Use barrier creams (Para-aminobenzoic acid).
	3. Contact dermatitis (more with chlorpromazine)	Hypersensitivity	Avoid contact Symptomatic (antihistaminics).
c. Haematological	1. Transient leucopenia and agranulocytosis (common with chlorpromazine & clozapine)	Idiosyncratic reaction	Stop drug Treat infection Add drug from another
	2. Rarely thrombocytopenic purpura, haemolytic anaemia, and Pancytopenia	Idiosyncratic reaction	Stop drug Add drug from another class (e.g. haloperidol)
	3. Blue-grey metallic discolouration.	Idiosyncratic reaction	Change drug
E. Metabolic and Endocrinal Side Effects	1. Galactorrhoea (with our without amenorrhea)	Dopaminergic blockade in hypothalamus leading to hyperprolactinemia.	Change drug, Amantadine Quitipine preferred
	2. Gynaecomastia	Dopaminergic blockade in hypothalamus leading to hyperprolactinemia.	Change drug, Amantadine
	3. Weight gain (not with molindone)	Not known	Dietary control, Exercise, change drug.
	4. Decreased libido, *Priapism* (especially with chlorpromazine, thioridazine).	Pituitary gonadotrophins and testosterone decrease Also anticholinergic, anti-adrenergic (1) effects.	Reduce dose or change drug.

Type	*Side effect*	*Mechanism of action*	*Management*
F. Cardiac	1. ECG changes 2. Sudden death	Anticholinergic effect Ventricular fibrillation Subictal discharges Respiratory depression	ECG monitoring Change drug Monitor vital signs. Start in low dose.
G. Ocular	1. Granular deposits in cornea 2. Pigmentary retinopathy (with *thioridazine*)	Not known (? Allergic) Not known Extrapyramidal signs	Careful follow up Change drug Not use thioridazine above 800 mg/day for prolonged period.
H. Pregnancy	1. First trimester	(Dopamine receptor blockade in foetus) increased fetal death Risk of teratogenesis	Avoid drug in first trimester (especially haloperidol) Use ECT
I. Antipsychotic Withdrawal Syndrome		Abrupt withdrawal results in increased dopaminergic noradrenergic, cholinergic and serotoninergic effects.	Gradually taper off, Continue antiparkinsonian for 2–3 days more.

Indications

(i) ***Psychiatric indications***

(a) ***Functional psychoses***

— *Schizophrenia* (control of acute attack as well as maintenance)
— *Mania*
— Schizoaffective psychosis (especially schizomania)
— Psychotic symptoms in major depression.
— Agitation in depression and other disorders.
— Infantile autism and Pervasive developmental disorder.

(b) ***Organic psychoses***

— Delirium (in low doses)
— Dementia (if there are psychotic features)
— *Postictal psychosis* (drugs with no or minimal effect on seizure threshold are preferred e.g. haloperidol, pimozide, trifluoperazine) or *interictal psychosis* (occurring in between attacks of epilepsy)
— Drug induced psychosis (e.g. haloperidol in amphetamine induced psychosis, pimozide in alcohol induced paranoid states etc.)
— Drug withdrawal states (e.g. haloperidol in delirium tremens etc.).

(c) ***Neuroses***

— Severe intractable anxiety (low doses)
— *Obsessive compulsive disorder (e.g. haloperidol in low doses).*
— *Monosymptomatic hypochondriasis (e.g. Pimozide).*
— *Secondary Hypochondriasis (if secondary to schizophrenia)*

(d) ***Attention deficit disorder with hyperactivity.***
(e) ***Tic disorder*** e.g. Gilles de la Tourette's syndrome (especially haloperidol)
(f) ***Conduct disorders*** (aggressive, destructive) in children.

(ii) ***Medical uses***

— Huntington's chorea (e.g. haloperidol)
— Nausea and vomiting, if central in origin.
— Intractable cough.
— To help patients to regain lost weight e.g. in anorexia nervosa.
— For relieving tension and emotional distress in physical illness.
— For relief of pain and distress in inoperable cases of secondary carcinoma.
— Preanaesthetic medication.
— Neuroleptanaesthesia (droperidol with fentanyl)
— *Hyperpyrexia*: to induce hypothermia.
— *Ecclampsia*: as a constituent of lytic cocktail (chlorpromazine + promethazine + pethidine)
— *Heat stroke.*
— *Pruritus.*

Contraindications and Precautions

They are given in Tables 3 and 4

Table 3: Contraindications and Special Precautions Required for Various Drugs

Drug group	*Contraindications*	*Special precautions*
I. Major tranquilizers	Depression, subcortical brain damage (Parkinsonism); impaired hepatic functions; blood dyscrasias, circulatory collapse, coma.	Use carefully in patient receiving other CNS depressant drugs, may lower seizure threshold, may disturb heat regulation, may produce hypotension, avoid in severe cardiovascular disorders. butyrophenones may reduce the effectiveness of oral anticoagulants.
II. Minor tranquilizers	Hypersensitivity, myasthenia, gravis, acute congestive glaucoma, pulmonary insufficiency, chronic psychoses.	Cardiorespiratory insufficiency, hepatic or renal dysfunction, with other CNS depressants.
III. Antidepressants	Hypersensitivity, heart block, narrow angle glaucoma, severe liver disease.	Cardiovascular disease, epilepsy, hyperthyroidism, glaucoma, urinary retention, renal or hepatic dysfunctions, use of other CNS2 depressants or anticholinergic drugs, suicidal tendencies.
IV. Stimulants	Heart disease, hypertension, tics, stereotypies, schizophrenia, anxiety states, hypersensitivity.	Hepatic disease, mentally retarded children, depression, chronic use, cerebrovascular or cardiac disease, glaucoma, urinary retention, anticoagulant therapy.
V. Lithium carbonate	Addison's disease; heart failure; severe renal insufficiency; thyroid dysfunction.	Dehydration or decreased salt intake, diuretic therapy, impaired renal function, pyrexia, electroconvulsive therapy.

Table 4: Side Effects of Atypical Antipsychotic (as compared to Haloperidol)

Effects	*Aripiprazole*	*Olanzapine*	*Risperidone*	*Ziprasidone*	*Quetiapine*	*Clozapine*	*Haloperidol*
EPS	0 to ±	± to +	0 to ±	0 to ±	0 to ±	0 to ±	+++
Dose related EPS	±	+	++	+	±	0	+++
Prolactin Elevation	±	±	++	±	±	±	+++
Anticholinergic Effects	±	+	±	±	±	+++	±
Hypertension	±	+	++	+	++	+++	+
Sedation	±	++	+	+	++	+++	+
QT Prolongation	0 to ±	±	±	++	±	++	±
Weight Gain	+	+++	++	+	+	+++	+
Total Cholesterol and triglycerides	↓	↑	↓	↓	↑	–	–

0 = None, ± = Minimal, + = Mild, ++ = Moderate, +++ = Severe, ↓ = Decrease, ↑= Increase.

Some conditions and the choice of antipsychotics are:

Disease/Disorder	Probable Preference
(a) ***Schizophrenia***	
Acute attack:	Any antipsychotic (chlorpromazine has additional sedative effect).
(b) ***Mania***	
Acute excitement:	Chlorpromazine (has additional sedative effect) Haloperidol (reduces psychomotor activity without causing much sedation). Risperidome or Olanzapine
(c) ***Schizoaffective***	
Schizomania:	Any (Chlorpromazine, haloperidol, Olanzapine)
Schizodepression:	Flupenthixol, Amoxapine, Olanzapine, Aripiprazole
(d) ***Infantile Autism***	Haloperidol
(e) ***Organic Psychoses***	
— Delirium, Dementia	Haloperidol, Pimozide etc. (because less sedation, clouding).
— Postictal or Interictal Psychosis	Haloperidol (Minimal no effect on seizure threshold)
— Drug induced or withdrawal states	Haloperidol, trifluoperidol, trifluoperazine (Minimal or no effect on seizure threshold)
(f) ***Neuroses***	
— Obsessive Compulsive Disorder	Haloperidol (sometimes)
— Monosymptomatic hypochondriasis	Pimozide
(g) ***Gille de la Tourette's syndrome***	Haloperidol, Pimozide
(h) ***To regain weight loss***	Phenothiazine, Risperidone, Clozapine, Olanzapine
(i) ***Pruritus***	Same

Thioridazine has *minimal* extrapyramidal side effects (clozapine has none whereas risperidone (up to 6 mg) and olanzapine (up to 10 mg) have fewer side effects) while the so-called high potency drugs such as haloperidol and thiothixene have fewest sedative and postural hypotension effects; safe (Butyrophenones) in hepatic impairments.

II. ANTIPARKINSONIAN DRUGS (Drugs used for treatment of Extrapyramidal syndromes)

- — Butyrophenones and piperazine derivatives are the most potent producers of extrapyramidal side effects.
- — Levo-dopa is not effective in drug induced parkinsonism and may induce psychiatric symptoms in about 15% of patients.

Classification, Indications and Dosages of various drugs

Refer to Table 5.

Table 5: Drugs for Treatment of Extrapyramidal Disorders

Generic name	*Starting dose*	*Uses*
I. Anticholingergic Drugs		
• Trihexyphenidyl	1 mg TID	Dyst, Akin, Park, Rabb, Proph
• Procyclidine	2.5 mg TID	— do —
• Orphenadrine	100 MG BID	— do —
	(60 mg I/V)	Dyst.
• Biperiden	2 mg TID	Dyst, Akin, Park, Rabb, Proph
	2 mg I.M./I.V.	Dyst.
• Benztropine	0.5 mg TID	Dyst, Akin, Park, Rabb, Proph
	1 mg I.M./I.V.	Dyst.
• Diphenhydramine	25 mg QID	Dyst, Akin, Park, Rabb, Proph
	25 mg I.M./I.V.	Dyst.
Promethazine	25 mg TID	Dyst, Akin, Park, Rabb, Proph
II. Dopamine Agonists		
• Amantadine	100 mg BID	Akin, Park, Rabb, Proph
• Bromocriptine	1.25 mg BID	NMS
III. Beta blockers		
Propranolol	20 mg TID	Akath
IV. Antidopaminergic		
Reserpine	1 mg	TD

Abbreviations: **Dyst: Dystonia; Akin : Akinesia; Park: Parkinsonism; Rabb: Rabbit Syndrome; Proph: Prophylactic Treatment; NMS: Neuroleptic Malignant Syndrome; TD: Tardive Dyskinesia.**

Side Effects

- — Reduce the serum levels of phenothiazines.
- — Acute organic syndromes (delirium) especially in elderly.
- — *Anticholinergic side effects* more when given with phenothiazines.
- — Excitement and euphoric effects.
- — My predispose to tardive dyskinesia or mask early symptoms.

III. ANTIDEPRESSANT DRUGS

They are known as *Mood Elevators* or *Thymo-leptics.*

The *first antidepressant* drug to be discovered was iproniazid (a monoamine oxidase inhibitor) by *Crane* (1957) and *Kline* (1958), imipramine (a tricyclic) was discovered by *Kuhn.*

Classification (See Table 6)

Table 6: Classification, Indications and Properties of Antidepressants

Class	*Examples of trade names*	*Avg. daily dose (mg)*	*Equivalent dose (to 75 mg imipramine)*	*Side Effects*			*Other side effects*	*Contra-indications*	*Inter-actions*
				Seda-tion	*Anti-choli-nergic*	*CVS*			
I. First Generation Tricyclics									
Imipramine	Depsonil, Depsol	75–300	75	++	+++	+++	Anticho-	Myocardial	Potentia-
Amitriptyline	Amitryn, Amitone	75–300	75	++++	++++	++++	linergic	infarction	tion of
Triimipramine	Surmontil	75–300	75	++	+++	++	Cardiac	Severe liver	alcohol &
Clomipramine	Anafranil, Clonil	75–300	75	++	+++	+++	arryth-	damage	barbiturates
Doxepin	Doxin, Doxetar	75–300	75–100	++++	++	+	mias,	Glaucoma	
Dothlepin	Prothiaden Dothip	75–300	75	+++	+++	++	Confu-	Urinary	
Nortriptyline	Sensival	75–250	75	+	++	+	sion,	obstruction	
							Drowsi-	Pregnancy	
							ness,		
							Weight	Clomipra-	
							gain,	mine with	
							Loss of	MAOI.	
							Epileptic		
							seizures		
							Blood		
							dyscrasias		
II. Second Generation									
a. Tricyclics Not Available/ in use									
b. Tetracyclics									
Mianserin	Depnon	30–120	20	+++	0	±	Seizures,	Myocardial	
							Bone	infarction	
							marrow	Pregnancy	
							depres-		
							sion		
c. Bicyclics Not Available/ in use									
d. Monocyclic/Unicyclic									
Bupropion	Bupron	300–450	300	a	0	0	Agitation,	Psychosis,	
							headache,	Seizures,	
							weight	Prolonged	
							loss,	Use.	
							seizures,		
							psychosis		
							GIT upset.		

Contd...

Class	Examples of trade names	Avg. daily dose (mg)	Equivalent dose (to 75 mg imipramine)	Side Effects Seda- tion	Anti- choli- nergic	CVS	Other side effects	Contra- indications	Inter- actions
e. Others									
Trazodone	Trazonil Trazolon	75–400	75	+++	0	0	Priapism	Epilepsy, Severe hepa- tic/renal disease	—
Flupenthixol	Fluanxol		0.75	++	+	±	Extra- pyrami- dal (rare)	Parkinson- ism, severe arteriosc- lerosis, delirium.	—
III. Third Generation									
Fluoxetine	Flunil, Prodep Fludac Dawnex Prodep	20–60 90–180 weekly	20	a	0	0	Nausea, insomnia weight loss, nervous- ness, head- ache	Hepatic/ renal disease, pregnancy MAOI	MAOI Tryptophan
Paroxetine	Paxidep, Panex,	20–40	20	a	0	0	,,	,,	Phenytoin
Sertraline	Xet, Serlift, Serta,	50–200	100	a	0	0	,,	,,	MAOI
Fluvoxamine	Fluvoxin,	50–300	150	a	0	0	—	,,	Phenytoin
Citalopram	Citopam,	20–60	20	+	±	0	— Warfarin	,,	Theophyllin
Escitalopram	Nexito, Feliz-S Rexipra	10–30	10	+	±	0	—	,,	
IV. Others									
Amoxapine	Demelox	150–300	150	++	+	+	Neurole- ptic mali- gnant syndrome seizures, tardive dyskinesia.	Parkinson- ism, hepatic/ renal disease, pregnancy MAOI	MAOI
Venlafaxine	Veniz, Ventab Venla, Venlor Venlift	75–375		+	±	±	Nausea Headache Sedation Dizziness Agitation Nervousness Dry mouth Impotence.	Hypertension Cardiac disease Seizures MAOI	MAOI
Tianeptine	Stablon	25–37.5		±	0	±	Nausea, Insomnia Dizziness Dry mouth	Hypersensitivity Hepatic disease MAOI	MAOI
Amineptine	Survector	100–400		0	0	0	,,	Hypersensitivity Huntington's Chorea closed angle glaucoma Pregnancy MI CRF, MAOI	MAOI

Contd...

Class	Examples of trade names	Avg. daily dose (mg)	Equivalent dose (to 75 mg imipramine)	Side Effects			Other side effects	Contra-indications	Inter-actions
				Seda-tion	Anti-choli-nergic	CVS			
Mirtazepine	Mirtaz Mirnite Mirt	15–45		+++	+	+	Nausea	Seizures Sedation Dizziness MAOI	MAOI
V. MAO Inhibitors									
a. Irreversible									
Nonselective	**Not Available/ in use**								
Selective	**Not Available/ in use**								
b. Reversible (selective)									
Selegiline (MAO-B inhibitor)		5–30	5	a	0	0	Insomnia	—	same as above
Moclobemide (MAOB-I)	(MAOB-I)	5–30	5	a	0	0	Jitteriness Insomnia	— —	Selectivity lost in higher doses
VI. Others									
Carbamaze-pine	Tegritol, Mazetol, Zen, Zeptol	600–1600	—	+++	0	±	Nausea, vomiting, diplopia, vertigo, tics, ny-stagmus, thyroid dysfunc-tion.	Hepatic in-sufficiency, Bone mar-row depres-sion. Pregnancy & lactation.	Potentiates sedative effects of alcohol & other drugs.
Lithium	Licab, Lithosun Intalith	600–1800	—	+	0	++	Hypothy-roidism, Diabetes insipidus cardiac toxicity GIT upset, myopathy.	Renal disease, Addison's disease, CHF	Diuretics (thiazides) Increases levels
Divalproex Sodium	Trend-XR, Divaa-OD, Dicorate - ER	750–1500	—	++	0	–	Nausea, Vomitting Weight gain Hair loss Skin react-ions	Hepatic in-sufficiency Bone marrow depression Pregnancy Lactation	Potentiates sedative effects of alcohol & other drugs

Abbreviations: — a–activating; 0–Absent; ± Probable; + mild; ++ moderate; +++ Severe; ++++ Very strong; —Not known.

Antidepressant drugs are classified on the basis of:

(a) ***Structure***

(*i*) *Tricyclic antidepressants*: Imipramine, Amitriptyline, Desipramine, Trimipramine, Nortriptyline, doxepin, clomipramine, dothiepin etc.

(*ii*) *Second Generation Antidepressants*

— Tetracyclics: Mianserin, Maprotiline

— Bicyclics: Zimelidine, Viloxazine

- Uni/Monocyclic — Bupropion
- Miscellaneous: — Amoxapine, trazodone, Nomifensine, Bupropion, Alprazolam.

(b) ***Biogenic amine reuptake blockade***

(*i*) *Both NE and 5-HT reuptake blockers*: Imipramine, Amitriptyline.

(*ii*) *Selective NE reuptake blockers*: Desipramine, Maprotiline.

(*iii*) *Selective 5-HT reuptake inhibitors (SSRIs)*: Clomipramine, Trazodone, Fluoxetine, Zimelidine, Paroxetine, Sertraline, Citalopram, Escitalopram.

(*iv*) *NE and dopamine reuptake inhibitors*: Nomifensine.

(*v*) *Weak or Non-reuptake*: Doxepine, Mianserin, Iprindole, Alprazolam.

(*vi*) *SNRIs*: Milnacipran

(c) ***Therapeutic uses***

(*i*) *Depression*

- Major depression (MDP depression, Endogenous depression) (along with ECT).
- Major depression with psychotic features or melancholia (with ECT's or anti-psychotics).
- Neurotic depression (with psychotherapy).
- Reactive depression (with psychotherapy)
- Atypical depression and unclassified depression (MAO inhibitors)
- Masked or latent depression.
- Depression in other psychiatric disorders (e.g. hysteria, schizophrenia, anxiety neurosis, hypochondriasis) and medical disorders (eg. malignancy, Cushing's syndrome etc.).

(*ii*) *Panic disorder* (with antianxiety drugs).

(*iii*) *Agoraphobia, Social Phobia, School Phobia* (MAO inhibitors).

(*iv*) *Obsessive Compulsive Disorder with or without depression* (Clomipramine and Fluoxetine are particularly helpful).

(*v*) *Enuresis* (with behaviour therapy).

(*vi*) *Chronic Pain*

(*vii*) *Attention deficit disorder* (in low doses, avoid tricyclics in children below 6 years of age because of cardiotoxicity).

(*viii*) *Bulimia nervosa*

(*ix*) *Migraine*

(*x*) *Peptic Ulcer disease*

(*xi*) *Cataplexy* (associated with narcolepsy).

(*xii*) *Miscellaneous*

- Abnormal grief reaction
- Trichotillomania (especially Clomipramine, Fluoxetine).
- Premenstrual and Menopausal syndromes
- Night terrors or somnambulism
- Cardiac arrythmias
- Tic Disorder

— Obesity (CNS stimulants)
— Depersonalization (CNS stimulants)
— Anorexia nervosa

Mechanism of Action

The exact mode of action of these drugs is not known. The main modes of action of these drugs are:

— Blocking the reuptake of norepinephrine and/or serotonin (5-HT) at nerve terminals, thus increasing their concentration at receptor site.
— Downregulation of beta adrenergic receptors.

Monoamine Oxidase Inhibitors (MAOI)

Mode of Action. The clinically used MAO inhibitors (MAOI) are irreversible inhibitors of MAO which metabolizes catecholamines and 5-HT. Harmine and Moclobemide inhibit MAO reversibly.

Adverse Effects

See Table 7.

Table 7: Adverse Effects of Antidepressants

Type	*Side effect*	*Mechanism of origin*	*Management*
I. Autonomic			
a. Anticholinergic	Dry mouth, Constipation, Urinary retention, Mydriasis, Cycloplegia, Precipitation of narrow angle glaucoma, Delirium,	Blockade of Muscarinic Cholinergic Receptors.	See Table above.
b. Antiadrenergic	Increased sweating.	Paradoxical effect	Don't use in elderly and patients with past history. Stop or change drug.
	Orthostatic hypotension Impaired ejaculation (Impotence)	Alpha 1 Adrenergic blockage.	See Table
	Hyperpyrexia and convulsions.	Interaction with tricyclics	Stop drug. Keep an interval of 10 days between 2 treatments. Supportive care.
c. Others	Priapism (with Trazodone)	Not known	Stop drug. Muscle relaxation or Surgery.
II. Cardiac	Quinidine like action (Increased AV conduction) Tachycardia ECG changes (Increased QT interval, flattening of T wave and S T segment) Arrythmias (in high doses, predisposed individuals) Direct myocardial depressant.	Anticholinergic	Use minimum dose, use newer safer drugs in elderly and in those with past history of cardiac problem.

Contd...

Type	Side effect	Mechanism of origin	Management
III. CNS	Sedation	α_1 adrenergic blockage	Start in low dose, Decrease dose or change it, give at night.
	Tremors & other extrapyramidal effects.	Not known	Decrease or change drug.
	Seizures	Decreases Seizure threshold	Decrease or change drug.
	Precipitation of Psychosis	Sympathomimetic	Stop drug, start in low dose.
	Precipitation of mania	Sympathomimetic	Stop drug, start in low dose.
	Jitteriness (early tricyclic syndrome)	Andrenergic	Tolerance occurs in 1–2 weeks.
	Withdrawal syndrome	Neuroadaptation	Slow withdrawal.
IV. Metabolic	Weight gain	Water retention, decreased activity due to illness and sedation of drug, increased appetite.	Exercise, Diet control, Change drug.
	Oedema (Occasionally)	Water retention	Reversible on stoppage.
V. Allergic Side Effects	Skin rashes	Hypersensitivity	Stop drug, antihistaminics
	Urticaria	"	Change drug,
	Cholestatic Jaundice	"	Benign course
	Agranulocytosis	"	Stop drug, Treat infection, Supportive care
	Pruritus	"	Stop drug, anthistaminics
	Photosensitivity	"	Avoid sun exposure. Use barrier creams (PABA).
VI. Specific Side Effects of MAO Inhibitors	Hypertensive crises (throbbing headache, palpitations, hyperpyrexia, convulsions, coma, death).	Interaction with tyramine containing foods (cheese, beer, red wine, chocolates etc). or indirectly acting sympathomimetic amines (e.g. ephedrine, amphetamine etc).	Dietary restriction and avoid use of sympathomimetic agents. Use alpha sympathetic blockers (e.g. phentolamine 5–10 mg I.V). Use safer, new reversible MAOI (selegiline, Moclobemide etc).
	Severe hepatic necrosis (uncommon) (with hydrazine derivatives)	Toxic (?), Hypersensitive	Stop drug, Supportive care, high mortality.
VII. Acute tricyclic overdose toxiicity (lethal dose 1–2 gm)	Hyperpyrexia, hypertension Convulsions Cardiac arrythmias Delirium, Coma.	Potentation of catecholamines. Anticholinegic	Gastric lavage, Cold sponges (for fever). Alpha adrenergic blockers (for hypertension). Diazepam (for convulsions). Propranolol (for arrythimias) Physostigmine (for anticholinergic side effects)
VIII. Acute withdrawal phenomenon.	Nausea, headache, restlessness, sweating, insomnia.	Psychological Dependence resulting in acute rebound	Gradual withdrawal. Avoid prolonged use.

IV. LITHIUM

It is the *lightest* of the alkali metals and was discovered by ***Arfuedson*** in 1817. Since ***J.F.J. Cade*** first reported the use of lithium as an antimanic drug in 1949.

Clinical Uses

(i) Mood (Affective) Disorders
- — Mania
- — Depression
- — Rapid Cyclers

(ii) Schizoaffective disorder

(iii) Alcoholism

(iv) Periodic Catatonia

(v) Uncontrolled Aggressive (Impulsive) Behaviour

(vi) Abnormal Mood Swings

(vii) *Other uses.* Migraine, premenstrual tension, tardive dyskinesia, thyroid disease (hyperthyroidism), *neutropenia, Felty's syndrome* and in conjunction with cytotoxic drugs, Kleine Levin syndrome (rarely) and Huntington's chorea.

Mechanism of Action

The exact mode of action is unknown.

Neurotransmitters

(i) *Synapses.* Increased presynaptic destruction of catecholamines, inhibits release of neurotransmitter, decreased sensitivity of postsynaptic receptor.

(ii) *Ions.* Influences sodium and calcium ion transfer across cell membranes.

(iii) *Cyclic AMP.* Inhibits prostaglandin E—stimulated cyclic AMP.

Cations and water

Stimulates exit of sodium from cells, probably by stimulating pump mechanism.

Cell membranes

Lithium may interact with both calcium and magnesium and increase cell membrane permeability.

Other actions

It restores diurnal rhythm of corticosteroids to normal in mania.

Lithium mediates Pineal gland stimulation resulting in serotoninergic fluorescence and increased melatonin (**"Tilak effect"**) (Parvathi Devi et al., 1982).

Pharmacokinetics

Lithium is an element with atomic weight 6.94 and atomic number 3. It is not metabolised in the body and excreted as such.

Dosage

Depending upon serum levels (about 600–1800 mg daily).

Therapeutic levels—0.6–1.2 meq/I (mOsm/L).

Prophylaxis—0.5–1.0 meq/I

Children and Elderly—0.4–0.8 meq/I.

Breast milk contains about ***one-third*** and saliva contains about ***twice*** that of serum lithium.

Absorption and Excretion

Lithium is administered as carbonate (most often), citrate or acetate salt.

— Absorption is rapid and is complete within 6–8 hours. Serum levels peak at 3–4 hours.

— Lithium is distributed in total body water.

— There is no protein binding and no metabolism. It is excreted unchanged by kidney.

— *Monitoring plasma levels.* The fasting blood sample is taken 12 hours after the last dose because of '*peaking*' of levels. Therapeutic and toxic ranges refer to the 'basic' level.

Preliminaries to Lithium Treatment

— A full blood count, plasma electrolytes and urea, creatinine clearance, and ECG and serum T4 and TSH levels are required.

— The lithium treatment is monitored initially by weekly serum levels (an estimation is done 5–7 days after any dosage change), followed by monthly (after stable levels are achieved) and then every 2–3 months.

— Thyroid function should be checked every six months.

— 24 hours urine volume should be done every 6 months and an ECG should be performed every year.

In acute administration, the gastrointestinal side effects are the ***commonest,*** though neurological side effects (especially tremors) are not uncommon. During long-term maintenance therapy, renal side effects are the ***commonest.***

Side Effects and Management

(See Table 8)

Table 8: Side Effects of Lithium and their Management

Side effect	*Management*
Gastrointestinal complaints	Give lithium after meals, give smaller doses more often, try slow-release preparation, lower the dosage
Tremor	Lower the dosage, give propranolol (40–100 mg/day), consider adding a benzodiazepine
Polyuria-diabetes insipidus	Try slow-release preparation, lower the dosage, add amiloride (5–10 mg/day), careful monitoring of lithium levels.
Acne	Benzotr peroxide (5–10%) topical solution, erythromycin (1.5–2%) topical solution
Muscular weakness, fasciculations, headaches	Usually resolve with first few weeks of treatment

Contd...

Side effect	*Management*
Hypothyroidism	Levothyroxine (0.05 mg qd), follow TSH level and increase of 0.2 mg qd as needed
T wave inversion	Benign, no treatment needed
Cardiac dysrhythmias	Usually must discontinue lithium
Psoriasis, alopecia areata	Dermatology consultation, reversible if lithium stopped
Weight gain	Difficult to treat, diet, may be partially reversible if lithium stopped
Edema	Consider spironolactone (50 mg PO qd); if severe, monitor lithium levels; resolves when lithium stopped
Leucocytosis	Benign, no treatment needed.

Drugs Interactions and Cautions

(i) ***Avoid*:** the lithium use with:
- — Diuretics
- — Low salt diet
- — Diarrhoea/vomiting
- — Obesity
- — Pregnancy (2nd and 3rd trimesters)
- — Dehydration
- — High grade fever
- — Parkinsonism

(ii) ***Use cautiously*:** in association with:
- — Major tranquilizers (especially haloperidol)
- — Thyroid disease
- — Renal insufficiency
- — Patients on Electroconvulsive therapy
- — Cardiac Patients

(iii) ***Contraindications***
- — Marked renal failure
- — Psoriasis
- — Myaesthenia gravis or myopathies
- — Addison's disease
- — Pregnancy (first trimester)/Lactation
- — Hypothyroidism
- — Impaired bone development
- — Acute myocardial infarction

V. HYPNOSEDATIVE DRUGS

A *hypnotic* drug is one which produces sleep resembling natural one.

A *sedative* is a drug that reduces excitement.

Classification

(i) *Urea derivatives*
- — Diureides: barbiturates
- — Related diureides: glutethimide, methyprylon

(ii) *Alcohols*: chloral hydrate, ethanol
(iii) *Aldehydes*: paraldehyde
(iv) *Acetylated carbinols*: Ethylchlorvynol
(v) *Benzodiazepines* and other tranquilizers
(vi) *Miscellaneous*: Methaqualone, antihistaminics, scopolamine
(vii) *Inorganic icons*: bromide.

The commonly used hypnosedative drugs are:

BENZODIAZEPINES

Sternbach discovered chlordiazepoxide in 1957.

Classification

The classification and properties of Benzodiazepines are given in Table 9.

Table 9: Classification and Properties of Benzodiazepines

Type	*Example*	*Half life (hours)*	*Peak time of effect (hours)*	*Oral dose (mg/d)*	*Hypnotic dose (mg HS)*
Long Acting					
	Chlordiazepoxide	5–30	2.4	10–100	10–30
	Diazepam	20–200	1.5–2	5–80	5–10
	Nitrazepan	20–60	2	5–20	5–20
	Flurazepam	40–250	1	15–60	15–60
Intermediate Acting					
	Oxazepam	5–15	1.4	15–120	15–30
	Lorazepam	10–20	1.4	20–10	1–2
	Alprazolam	6–20	1.2	0.5–6	0.5–1.0
	Estazolam	15–20	1.2	1–2	0.5–1.0
Short Acting	Triazolam	1.5–5	2	0.25–1.0	0.25–0.5

Therapeutic uses and specificity of benzodiazepines

- — Acute and chronic anxiety—chlordiazepoxide.
- — Mixed anxiety-depression states-Alprazolam
- — Status epilepticus—Diazepam
- — Myoclonic and petitimal seizures—clonazepam
- — Neuromuscular disorders e.g. cerebral palsy and stiffman syndrome—Diazepam
- — Insomnia—Nitrazepam, flurazepam, temazepam, triazolam.

- Alcohol withdrawal syndrome—Chlorazepate, chlordiazepoxide, diazepam and oxazepam.
- Absence seizures and other type of childhood seizures—Clonazepam
- Sedation—Anaesthesia—Midazolam.
- Panic disorder with phobias—Alprazolam, oxazepam, lorazepam.
- Anxiety in patients with hepatic impairment—Oxazepam, lorazepam.

Mechanism of Action

The exact mode of action of benzodiazepines is not known.

Benzodiazepine Receptors (BZs Receptors)

(i) *Central receptors*

- ***BZ_1 type.*** These are predominant in cerebellum and responsible for anxiolytic action.
- ***BZ_2 type.*** They are mainly responsible for anticonvulsant and hypnotic effects, predominant in cerebral cortex.

(ii) *Peripheral receptors* ("*Acceptors*") are found in mast cells, liver, heart, platelets, lymphocytes etc.

Mode of Action

Benzodiazepines are believed to *potentiate GABA* (an inhibitory neurotransmitter) activity by increasing the *frequency* of chloride channel opening (whereas barbiturates potentiate GABA activity by simply increasing the time that the chloride channel remain open).

Contraindications

- *Respiratory insufficiency.* Administer Benzodiazepines with care in elderly and in patients with limited pulmonary reserve.
- *Hepatic failure.* Oxazepam and lorazepam are safer.
- *Obstetrics.* They produce ***"Floppy infant" syndrome*** manifested by hypotonia, lethargy and sucking difficulties in newborns.
- *Pregnancy and lactation.* May increase the risk of congenital malformations during first trimester of pregnancy (e.g. diazepam may increase the risk of *cleft palate and lip* in babies).
- *Renal insufficiency.*
- Acute intermittent porphyria e.g. chlordiazepoxide
- *Tartrazine insensitivity.*
- *Paradoxical reactions*: In hyperactive, aggressive children, excitement, stimulation or acute range have been reported.
- *Analgesics*: With benzodiazepines, the dose of narcotic analgesics should be reduced to one-third.
- *Shock, coma and acute alcohol intoxication.*
- *Acute narrow angle glaucoma.* Alprazolam and chlordiazepoxide are avoided.

Adverse Effects

The side effects of BZs include drowsiness, lethargy, impaired psychomotor performance, gastric upset (nausea, vomiting, diarrhoea, epigastric pain), blurring of vision, bodyaches, impotence, urinary incontinence, ataxia in high doses), retrograde and antegrade amnesia, disinhibited behaviour, dependence and withdrawal syndrome, cross tolerance with barbiturates and alcohol and coma. Benzodiazepines may produce nightmares, paradoxical delirium, confusion, depression, aggression, hostile behaviour, metallic taste and headaches.

NON-BENZODIAZEPINE ANXIOLYTICS

— *Pyrazopyridines*: Etazolate and cartazolate increase the binding ability of BZ receptors.

— *Zopiclone*: A pyrollopyrazine, has very high affinity for central BZ receptors.

— *Atypical compounds—Buspirone,* an azaspiro-decanedione, is anxiolytic which acts without interacting with BZ receptors. It is a potent dopamine stimulant which indicates the role of dopamine in the aetiology of anxiety. Buspirone is a selective dopamine autoreceptor antagonist. It lacks hypnotic, anticonvulsant and muscle relaxing properties, hence anxioselective, causes less sedation, no dependence and no withdrawal syndrome. It does not potentiate the effects of alcohol. It is given in the dose of 10-30 mg/day (in divided doses). It is completely metabolised and its half life is 2 to 3 hours.

Side Effects

It includes dizziness, headache, light-headedness and diarrhoea.

VI. CNS STIMULANTS

Indications

(*i*) Hyperkinetic syndrome

(*ii*) Narcolepsy

(*iii*) Enuresis

(*iv*) Obesity

(*v*) Depression

21

Electroconvulsive Therapy

(ECT, also Electroshock or Shock Therapy or Electroplexy therapy)

Indications

The main indications of ECT include

(i) ***Depressive illness*:** ECT is effective treatment in severe depressive illness especially with somatic features (i.e., sleep disturbance, loss of appetite and weight, psychomotor retardation etc.) and psychotic symptoms (e.g. delusions, feeling of guilt, suicidal tendencies etc.). Other indications in depression are:

- severe depression with suicidal risk.
- depressive Stupor.
- depression with marked retardation.
- severe puerperal depression
- depressive illness with nihilistic or paranoid delusions.
- patients with Schizodepression.
- failure to respond to an adequate course of antidepressant.
- inability to tolerate side effects of antidepressants.
- in the elderly where ECT may be safer than drugs.
- inability to take drugs e.g. depression in first trimester of pregnancy, depression in physical illness e.g. liver or renal failure.

The combination of ECT and antidepressants has been found to produce fewer relapses at follow up.

(ii) ***Schizophrenia*:** ECT produces greater early symptomatic relief than the neuroleptics but when both are combined, the benefit is maximum. The main indications of ECT in schizophrenia are:

— excitement (secondary to catatonia or delusions)
— stupor (catatonic)
— acute schizophrenic episode
— intolerable or resistant to drugs
— puerperal schizophrenia
— schizophrenic episode in first trimester of pregnancy
— schizophrenia in the presence of chronic physical illness when drugs are contraindicated
— depression in the schizophrenic illness (e.g. at onset or in residual phase)

(iii) ***Mania*:** The main indications of ECT in mania are:
— excited or uncooperative behaviour
— bipolar mood disorder with mixed features
— bipolar mood disorder-rapid cyclers
— others-mania in first trimester of pregnancy, puerperal mania, schizomania.

(iv) ***Postpartum psychosis*:** Some reports indicate that ECT is the treatment of choice in Puerperal psychosis (especially depressive or bipolar type or mixed type).

(v) ***Schizoaffective disorders*:** ECT is an effective mode of treatment in patients with schizodepression.

(vi) ***Psychosis in first trimester of pregnancy*:** When drugs cannot be given, ECT is treatment of choice. Pregnancy is not at all a contraindication.

(vii) It is also useful in psychotic patients who may have underlying mental retardation, epilepsy, organic mental disorders and delirium tremens. ECT has also been tried in anorexia nervosa, neurodermatitis, phantom limb and trigeminal neuralgia.

Side Effects

(i) ***Headache, bodyaches and vomiting*** (due to temporary increase in intracranial pressure and myalgias)

(ii) ***Confusion*:** It is usually slight and temporary. Prolonged confusion can be due to underlying organic illness or when duration of current or voltage was more.

(iii) ***Amnesia*:** (both retrograde and anterograde). Memory impairment that occurs with ECT is highly variable. The memory loss is short-term and may last from a few days to few months (from 9 days to 9 months). The memory loss is believed to be due to neuronal hypoxia during seizure. It can be minimised by:
— using unilateral ECT.
— oxygenation before and after seizure.
— recall of major events or routines before ECT.
— giving individualised minimal voltage and current.

— giving minimum number of ECT's with proper spacing.

— usse of brief pulse stimulation than sine-wave stimulation.

— avoiding in elderly if hypertensive or diabetic.

(iv) ***Other*:** Rare complications like fractures (of thoracic spine and long bones e.g. femur, humerus are *commonest*), dislocations (of temporomandibular joint, shoulder, wrist joint are common) and fat embolism.

Mortality

The details are rare after ECT. It has been estimated as 1 to 10 per lac ECT treatments.

Reasons of Death

— Coronary thrombosis and cardiac arrest occurs due to vagal inhibition.

— Sometimes aspiration of gastric contents and respiratory depression may cause death.

Contraindications

(*a*) ***Absolute*:** *Raised intracranial pressure*

(*b*) ***Relative***

(*i*) *Contraindications to ECT*: They are

— myocardial infarct in previous 2 years.

— other cardiac disease including arrythmias.

— pulmonary disease (cavitating tuberculosis, pneumonia, bronchial asthma etc.).

— others e.g. fractures, myopathies, fever, dehydration, glaucoma, retinal detachment, CHF, angina, thrombophlebitis etc.

(*ii*) *Contraindications to Anaesthesia and Other Agents*: The contraindications consist of those to *atropine* (glaucoma, arrythmias, prostatic hypertrophy etc.), *succinylcholine* (myasthenia gravis, myopathies, those with family history of pseudocholinesterase deficiency etc.) and *barbiturates* (porphyria, hepatic disorders, respiratory dysfunctions etc.). ECT can be safely given in women during any stage of pregnancy, patients with cardiac pacemakers and to epileptics.

(*iii*) *Psychological Contraindications*

— phobic neurosis

— depersonalisation syndrome

— obsessive compulsive disorder

— hysterical neurosis

— primary hypochondriacal neurosis.

Mode of Action of ECT

The exact mode of action is *unknown*. The various hypotheses to explain its effectiveness:

(*i*) The repeated rapid induction of unconsciousness.

(*ii*) The passage of electricity across the brain.

(*iii*) The induction of bilateral grand mal epileptic seizure and neurotransmitters balance.

(*iv*) The administration of an anaesthetic drug, a muscle relaxant and sometimes atropine.

(*v*) Considerable medical and nursing attention.

(*vi*) A varied set of attitudes and expectations on the part of the patient and the family.

It appears that the electricity and/or epileptic fit are necessary for ECT to exert its full effect but that in non-deluded depressed patients, other factors such as 1, 5 or 6 above may play a part.

Administration of ECT

(*i*) ***Explanation to the patient*:** Fears and fantasies about the treatment are often allayed by the facts and rarely made worse. Fasting of 6–12 hours is needed.

(*ii*) ***Consent*:** Informed or real consent requires an understanding of the nature, purpose and likely consequences of a treatment.

(*iii*) ***Fully physical examination*:** including Fundus, ECG etc.

(*iv*) ***Testing for cerebral dominance*:** This should be carried out routinely on all patients who are to receive unilateral ECT.

Application

(*i*) ***Premedication and Anaesthetic*:**

— ***Anaesthetic agent*:** Methohexitone sodium (0.2% solution in 5% dextrose) is the drug of choice of ECT anaesthesia. In India, thiopentone is commonly used (0.25 gm in 10 C.C) i.e., 5–10 C.C. of 2.5% solution and not 5.0% solution, given over 20 to 25 seconds. Its main indication is to produce anaesthesia.

— ***Muscle relaxant*:** A muscle relaxant i.e., succinylcholine is used. Its dose is about 30 mg (0.5 mg/kg) intravenously (more dose may be required in patients with rheumatoid arthritis i.e., 50 mg). The dose should be adjusted as not to abolish all signs of convulsion.

— ***Atropine*:** Atropine should not be given routinely by subcutaneous or intramuscular route as a premedication unless there has been a previous problem with excess salivation, bronchial secretion. The main reason to give atropine is to block the vagus nerve and so protect the heart from bradycardia and arrythmias. Vagal blockade can be achieved by giving atropine sulphate 0.6 to 1 mg intravenously at the time of giving anaesthetic agent.

— ***Oxygenation*:** Oxygen should be given before and after the period of succinylcholine-induced apnoea.

(*ii*) ***The Current*:**

- After placing a mouth gag in patient's mouth (to prevent tongue bite or lip bite), an electrical stimulus (70 to 120 volts which varies from patient to patient) for 0.2 to 0.6 seconds is passed. It is important to clean the scalp with normal saline or jelly to decrease scalp's resistance.

How to know whether a fit has occurred

- The most reliable is *EEG monitoring*.
- Isolate one arm by inflating a blood pressure cuff to above systolic blood pressure before the muscle relaxant is given. It will not become paralysed by the relaxant and will show twitchings. *When Unilateral ECT is used, apply cuff to the ipsilateral forearm.*
- *Others.* The other signs such as bilateral plantar extensor, reaction of pupil (if constriction and then slow dilatation) are not reliable (especially if unilateral ECT is used).

Number and Frequency of Treatments

A set number of treatments should not be prescribed. Assess the patient after each ECT.

Most patients need 4 to 12 ECTs.

22

Psychological Methods of Treatment

I. PSYCHOTHERAPY

Definition

Psychotherapy is the development of a trusting relationship, which allows free communication and leads to understanding, integration and acceptance of self.

Common Features of Psychotherapies

— An intense, emotionally charged relationship, with a person or group.

— A rationale or myth explaining the distress and methods of dealing with it.

— Provision of new information about the future, the source of the problem and possible alternatives which hold a hope of relief.

— Non-specific methods of boosting self-esteem.

— Provision of success experiences.

— Facilitation of emotional arousal

— It takes place in a locale designated as a place of healing.

Types of Psychotherapy

Psychotherapies are classified according to:

(a) ***Depth of probing in the unconscious mind***

— Superficial or short-term (also known as *supportive psychotherapy*).

— Deep or long-term (also known as *analytic psychotherapy*).

— Educative (also known as *counselling*).

(b) ***Number of patients treated in any therapeutic session***

(*i*) Individual Psychotherapy.

(*ii*) Group Psychotherapy.
(*iii*) Family Therapy.

(c) ***Theoretical formulations used in psychotherapy***

(*i*) *Supportive*. Which provide support, guidance, advice and reassurance.
(*ii*) *Re-educative.* Which attempt to teach the individual new patterns of behaviour and social functioning.
(*iii*) *Reconstructive.* Which aim to dismantle and rebuilt a new personality.

Unwanted Effects of Psychotherapy

— Patients may become excessively dependent on therapy or the therapist.
— Intensive psychotherapy may be distressing to the patient and result in exacerbation of symptoms and deterioration in relationships.
— Disorders for which physical treatments would be more appropriate e.g. psychotic states or physical illness presenting with mental symptoms, may be missed.
— Ineffective psychotherapy wastes time and money, and damages patients' morale.

Contraindications

— Psychotic patients with severe behaviour disturbances like excitement.
— Organic psychoses (in acute phase).
— Patients who are unmotivated and unwilling to accept it.
— Group psychotherapy in hysteria, hypochondriasis etc.
— Patients who are unlikely to respond e.g. personality disorders (especially antisocial type), malingering etc.

Commonly used Psychotherapies are

I. Supportive Psychotherapy

It is a form of psychological treatment given to patients with chronic and disabling psychiatric conditions for whom basic change is not seen as a realistic goal.

Indications (Selection Criteria)

The main indications of supportive psychotherapy include:

(*i*) The 'healthy' individual faced with overwhelming stress or crises—particularly in the face of traumas or disasters.
(*ii*) *Patient with ego-deficits.*
(*iii*) *Other Indications* e.g. *Alexithymic* patients (those with a striking inability to find words to describe other emotions and a tendency to describe endless situational details or symptoms instead of feelings), Passive *patients* those who derive *significant practical benefit,* patients who are *able to relate to the therapist,* have past history of good interpersonal relationship, word and educational performance and use leisure time are better suited for supportive therapy.

Table 1: Comparison of Psychologic Techniques

		Counselling	*Psychotherapy*
(*i*)	Duration	Brief (< 6 months)	Brief or long-term
(*ii*)	Clinical skill required	Primary care	Specialised mental health training
(*iii*)	Setting	Informal	Structured
(*iv*)	Use of transference	No	Yes
(*v*)	Strategy	Supportive	Insight directed
(*vi*)	Patient's ego strength	Intact	Threatened or intact
(*vii*)	Use of medication	No	May be necessary

Techniques

(*i*) The Interview
(*ii*) Reassurance
(*iii*) Explanation (Interpretative Comments)
(*iv*) Guidance and Suggestion
(*v*) Ventilation

II. Client-centred Psychotherapy

It was borrowed from the ideas of Carl Rogers (1951). The term 'client' rather than 'patient' has been adopted. The *main conditions* include:

- The client is experiencing at least a vague incongruence which causes him to be anxious.
- The therapist is congruent (or genuine or real) in the relationship.
- The therapist is experiencing a prizing, caring or acceptable attitude toward the patient.
- The therapist is experiencing an accurately sensitive understanding of the client's internal frame of reference.
- The client perceives to some minimal degree of realness, the caring and the understanding of the therapist.

The following attitudes are deemed to be most important for the success of client-centred therapy:

(i) Genuineness (or Congruence)
(ii) Unconditioned Positive Regard
(iii) Accurate Empathy

III. Cognitive Psychotherapy

Definition

It is a group of psychological treatments which share the aim of bringing about improvement in psychiatric disorder by altering maladaptive thinking.

It was developed by *Aaron T. Beck and his colleagues.* See Table 2.

Table 2: Cognitive Processing Errors

1. **Emotional reasoning:** A conclusion or inference based on an emotional state; i.e., "I feel this way; therefore, I *am* this way."
2. **Overgeneralisation:** Evidence drawn from one experience or a small set of experiences to reach an unwarranted conclusion with far-reaching implications.
3. **Catastrophic thinking:** An extreme example of overgeneralisation, in which the impact of a clearly negative event or experience is amplified to extreme proportions; e.g., "If I have a panic attack I will lose *all* control and go crazy (or die)."
4. **All-or-none (black-or-white; absolutistic) thinking:** An unnecessary division of complex or continuous outcomes into polarised extremes; e.g., "Either I am a success at this, or I'm a total failure."
5. **Shoulds and musts:** Imperative statements about self that dictate rigid standards or reflect an unrealistic degree of presumed control over external events.
6. **Negative predictions:** Use of pessimism or earlier experiences of failure to prematurely or inappropriately predict failure in a new situation; also known as "fortune telling."
7. **Mind reading:** Negatively toned inferences about the thoughts, intentions, or motives of another person.
8. **Labelling:** An undesirable characteristic of a person or event is made definitive of that person or event;' e.g., "Because I *failed* to be selected for ballet, I am a *failure.*"
9. **Personalisation:** Interpretation of an event, situation, or behaviour as salient or personally indicative of a negative aspect of self.
10. **Selective negative focus (selective abstraction):** Focusing on undesirable or negative events, memories, or implications at the expense of recalling or identifying other, more neutral or positive information. In fact, positive information may be ignored or disqualified as irrelevant, atypical, or trivial.
11. **Cognitive avoidance:** Unpleasant thoughts, feelings, or events are misperceived as overwhelming and/ or insurmountable and are actively suppressed or avoided.
12. **Somatic (mis) focus:** The predisposition to interpret internal stimuli (e.g., heart rate, palpitations, shortness of breath, dizziness, or tingling) as *definite* indications of impending catastrophic events (i.e., heart attack, suffocation, collapse, etc.).

Classification of cognitive therapy

(*i*) *Techniques Intended to Interrupt Cognition.* These aim to stop sequence of intrusive thoughts in the hope that thoughts will not start again immediately. Because intrusive thoughts are difficult to arrest, another technique *'thought stopping'* is used in which a sudden, intense but short lived distraction is given, the common being just shouting aloud to stop, then the patient repeats silently; the other method used is by applying painful stimuli.

(*ii*) *Techniques intended to Counterbalance Cognitions.*

(*iii*) *Techniques intended to alter cognitions.*

(*iv*) *Techniques for problem solving.*

Applications

Depressive Disorder

(*i*) Intrusive Thoughts e.g. low self-regard, self-criticism, self-blame.

(*ii*) ***Cognitive disorders***

- *Arbitrary inference*, e.g. patient sees a friend in the street who fails to acknowledge him, he thinks his friends do not like him any more.
- *Selective abstraction,* e.g. teacher in his class sees two students bored, feels his class is not liked by all the students.
- *Overgeneralisation* e.g. A mother spoiling a dish feels that she is a bad mother.
- *Magnification or minimisation*, e.g. person commits an unimportant error, thinks that his employer has noticed it and he will be terminated of the job (Magnification);
- A depressed patient makes a great effort to help his friend in trouble, yet fails to accept that he is doing his best (Minimisation).

Other Indications

- Anxiety neurosis.
- Phobias (especially agoraphobia)
- Eating disorders (e.g. anorexia nervosa, bulimia nervosa)
- Problems involving bereavement, divorce, redundancy at work etc.
- Alcoholic patient.

Technique

See Table 3.

Table 3: Structure of a Typical Cognitive Therapy Session

1. Mood check

   Examine symptom severity scores from a questionnaire such as the Beck Depression Inventory.
2. Set the agenda
3. Weekly items
   - Review of events since last session
   - Feedback on reactions to previous session and review of key points
   - Homework review
4. Today's major topic
5. Set homework for next week
6. Summarise key points of today's session
7. Feedback on reactions to today's session

IV. Marital Therapy: In marital therapy, treatment is given to both partners in a marriage. The term *'couple therapy'* is sometimes used.

Definition: Marital therapy, the treatment of marital relationship, refers to a broad range of treatment modalities that attempt to modify the marital relationship with the goal of enhancing marital satisfaction or correcting marital dysfunction. *Marital therapy* differs from *marital counselling* on theoretical and technical basis i.e., marital therapy employ varied, extensive assessment techniques and utilise the systematic knowledge of personality, learning and communicational systems theory to achieve the goal whereas marriage counselling includes a very broad range of disharmony.

Indications

— Presence of overt marital conflicts that result in recognisable suffering of both spouses.
— Presence of covert marital disorder in form of symptomatology or dysfunction in one of the spouses or the children.
— Poor communication problem.
— Extramarital relationships.
— When individual therapy has failed or is unlikely to succeed due to poor motivation or limited ability to communicate with the therapist.
— Eruption of symptoms in a family member coincides with the outbreak of marital conflicts.
— When there is danger of marital instability due to improvement of a mentally ill patient or when the healthy "spouse" develops symptoms.

Contraindications

— Premature exposure of the spouses to marital secrets such as homosexuality, criminal involvement.
— If the spouses use the sessions consistently to attack each other.
— Lack of commitment to continuation of the marriage.

V. Family Therapy

History: *Nathan Ackerman* developed the idea of family therapy as a result of his experience in the psychotherapy of children.

General principles: The essential features of family therapy are the following concepts.

— The parts of the family are interrelated.
— One part of the family cannot be fully understood in isolation from the rest of the system.
— Family functioning cannot be fully understood by simply understanding each of the parts.
— A family's structure and organisation are important factors determining the behaviour of family members.
— Transactional pattern of the family system shapes the behaviour of family members.

Indications

Family therapy has been used in all types of psychiatric problems including the psychoses, reactive depression, anxiety neurosis, psychosomatic illness, substance abuse and various childhood psychiatric problems.

Contraindications

- Lack of adequately trained therapist (only true contraindication)
- Poor motivation
- Fixed character pathology e.g. lying, physical violence.
- Extreme secrecy.

VI. Group Psychotherapy: In 1919, *L. Cody Marsh* applied the group method of treatment to institutionalised mental patients.

Definition

Group psychotherapy is a form of treatment in which carefully selected emotionally ill persons are placed into group, guided by a trained therapist for the purpose of changing the maladaptive behaviour of the individual member.

Patient Selection

Inclusion Criteria

- Ability to perform the group task.
- Problem areas compatible with goals of group.
- Motivation to change.
- Patients with authority anxiety (especially adolescents).
- Patients using defense mechanisms of projection, repression, denial, suppression, transference reactions.

Exclusion Criteria

- Marked incompatibility with group norms for acceptable behaviour.
- Inability to tolerate group setting (Peer anxiety).
- Severe incompatibility with one or more of the other members.
- Tendency to assume deviant role.

Special Indications

The main uses of group therapy include

- Schizophrenia
- Mood Disorders
- Paranoid States
- Neuroses (Anxiety Neurosis, Phobic Disorders)

- Personality Disorders (schizotypal, borderline, schizoid, passive-aggressive, dependent, avoidant, rarely antisocial type).
- Disorders of Impulse Control
- Adolescent Disorders
- Other Disorders e.g. homosexual conflict disorder, transvestism, gender identity disorder, alcoholism and other substance abuse disorders, juvenile delinquency etc.

II. BEHAVIOUR THERAPIES

A. BEHAVIOUR THERAPY

Definition

It is the *systematic application of principles of learning* to the analysis and treatment of disorders of behaviour.

Learning

It is defined as any *relatively permanent change* in behaviour which occurs as a result of practice or experience.

Behaviour

Strictly speaking behaviour refers to the organism's skeletal muscle activity in humans, both what they do (motor behaviour) and what they say (verbal behaviour).

I. Classical Conditioning (CC)

Ivan P. Pavlov (1849–1936): *Pavlovian or respondent* conditioning. The essential operation in classical conditioning (CC) is a pairing of two stimuli.

A neutral conditioned stimulus (CS) is paired with an unconditioned stimulus (US) that evokes an unconditioned response (UR). As a result of this pairing, the previously neutral conditioned stimulus begins to call forth a response similar to that evoked by the unconditioned stimulus. After learning, when the conditioned stimulus produces the response, the response is called a *conditioned response* (CR).

Terms used

Extinction: The weakening of a conditioned response occurs in CC when the CS is repeatedly presented without the US.

Stimulus generalisation: Tendency to give CR to stimulus which are similar in some way to the CS but which have never been paired with the passage of time.

Discrimination: Process of learning to make one response to one stimulus and another response or no response to another stimulus.

Classical conditioning (CC): With respect to human behaviour, CC seems to play a large role in the formation of conditioned emotional responses the conditioning of emotional states to previously neutral stimuli.

II. Operant Conditioning (OC)

(By *B.F. Skinner,* 1904–1990). In OC, a reinforcer in any stimulus or event which when produced by a response, makes that response more likely to occur in future.

The major principle of OC is that if a reinforcement is contingent upon a certain response, that response will become more likely to occur.

Terms used

***Shaping*:** Process of learning a complex response by first learning a number of similar responses which are steps leading to the complex response.

***Extinction*:** In OC, extinction of learned behaviour—a decrease in the likelihood of occurrence of the behaviour is produced by omitting reinforcement following the behaviour.

***Stimulus generalisation*:** Same as in CC.

***Discrimination*:** Develops in OC when differences in the reinforcement of a response accompany different stimuli.

***Continuous reinforcement*:** Reinforcement follows every occurrence of a particular response called *continuous reinforcement.*

***Primary reinforcer*:** In OC, it is one which is effective for an untrained organism: no special previous training is needed for it to be effective.

***Secondary reinforcer*:** Is a learned reinforcer; stimuli become secondary reinforcer; stimuli which become paired with primary reinforcers.

***Positive reinforcer*:** It is stimulus or event which increases the likelihood of a response when it terminates or ends, following a response.

— Praise is the easiest one

— Reinforce the reinforcers.

***Negative reinforcers*:** Are noxious or unpleasant, stimuli or events which terminate when contingent upon the appropriate response being made.

***Escape learning*:** The acquisition of responses which terminate noxious stimulation—is based on negative reinforcement.

***Punisher*:** In contrast to negative reinforcement, a punisher is a noxious stimulus that is produced when a particular response is made. Punishers decrease the likelihood that a response will be made and thus involved in learning what not to do.

Classical Conditioning versus Operant Conditioning (CC versus OC)

— In OC reinforcement is contingent on what the learner does while in CC reinforcement is defined as the pairing of the conditioned and unconditioned stimuli and is not contingent on the occurrence of a particular response.

— The responses which are learned in CC are stereotyped, reflex like ones which are elicitated by the unconditioned stimulus while in OC response is voluntary.
— In CC—consequences of behaviour are relatively unimportant while in OC they are important.

III. Cognitive Learning

Cognitive learning is learning in which without explicit reinforcement, there is a change in the way information is processed as a result of some experience a person or animal has had.

Behaviour Therapy versus Psychoanalysis

— Behaviour therapy asserts that the symptom is the illness and not that there is any underlying process or illness of which the symptoms are merely superficial manifestations.
— Behaviour therapy is applicable to unwilling patients.

Principles of Behaviour Therapy

— Close observation of behaviour.
— Concentration on symptoms as the target for therapy.
— General reliance on principles of learning.
— An empirical approach to innovation.
— A commitment to objective evaluation of efficacy.

Indications of Behaviour Therapy (BT)

It is a *treatment of choice* in
— Phobias
— Compulsions
— Nocturnal enuresis
— Social anxiety states
— Sexual dysfunctions
— Tension headaches
— Tics
— Obesity
— Anorexia nervosa.

Also used to modify
— Maladaptive habits
— Sexual role disturbances
— Psychosomatic reactions
— Smoking
— Drinking

Contraindications of BT

Those psychiatric disorders in which symptomatology in *acute, pervasive* or non circumscribed and in which triggering environmental events or external reinforcement are *not obvious* or capable of definition.

Techniques

(a) ***Systematic desensitisation*** **(Wolpe, 1958):** It is based on the principle of *reciprocal inhibition,* which holds that prior establishment of an appetitive physiological response can prove capable of blocking a conditioned avoidance response.

Systematic desensitisation as given by Wolpe (1958) involve the following *three stages*:

(*i*) Training the patient to relax

(*ii*) Constructing with the patient a hierarchy of anxiety-arousing situation.

(*iii*) Presenting phobic items form the hierarchy (a sequence of phobic stimuli in an increasing order) in a graded way, whilst the patient inhibits the anxiety by relaxation.

In phobic neurosis, it is extensively used.

(b) ***Flooding*****:** Flooding involves exposing patients to a phobic object or situation in a non-graded manner with no attempt to reduce anxiety. Unlike systematic desensitisation, no prior relaxation techniques are taught to the patient and it is usually given in a non-graded manner or in reverse hierarchy (starting from most phobic to least phobic stimulus). It can be conducted in imagination (*Implosion*) or in vivo.

Flooding is avoided in patients with cardiovascular disorders or uncooperative patients or those who continue to have panic attacks.

(c) ***Shaping*****:** The successive approximations to the required behaviour with contingent positive reinforcement. It is useful in many other types of situations e.g. rehabilitation of physically handicapped children, children with neurotic behaviour or autism, etc.

(d) ***Modelling*** **(Bandura et al., 1969):** It refers to the acquisition of new behaviours by the process of imitation. In this form of treatment, the patient observes someone else (may be the therapist) carrying out an action which the patient currently finds difficult to perform.

(e) ***Response prevention and restraint*****:** When combined with flooding, it is the *treatment of choice in obsessive compulsive neurosis.* The technique involves exposing the patient to a contaminating object, such as soiled towel and subsequently preventing him from carrying out his usual cleansing ritual. *Thought stopping* is sometimes used in the control of obsessional thoughts by arranging a sudden intrusion.

(f) ***Aversion*****:** It involves producing an unpleasant sensation in the patient, usually by inflicting pain in association with a stimulus.

Aversion therapy has been used for alcoholism and sexual perversions.

(g) ***Self-control techniques***

- *Self Monitoring*: It refers to keeping daily records of the problem behaviour and the circumstances in which it appears, e.g. a patient with bulimia nervosa.
- *Self evaluation*: It refers to making records of progress and this also helps to bring about change.

***Premack principle*:** Any frequently performed piece of behaviour can be used as a positive reinforcer of the desired behaviour.

(h) ***Contingency management*:** This group of procedure is based on the principle that if behaviour persists, it is being reinforced by certain of its consequences and if these consequences can be altered, the behaviour should change.

***Token Economy*:** When reinforcement is mainly by tokens to be exchanged for privileges.

(i) ***Assertiveness training* (Wolpe, 1958):** Used in chronically depressed, socially anxious and inhibited in the expression of warm feelings of anger.

(j) ***Negative practice* (Dunlap, 1932):** Some problems e.g. tics, stammering, thumb-sucking, nailbiting etc. can be reduced when the patient deliberately repeats the behaviour.

(k) ***Biofeedback*:** Discussed later in this chapter.

(l) ***Social skills training*:** Discussed later in this chapter.

(m) ***Cognitive therapy*:** Discussed with psychotherapy.

(n) ***Hypnosis and abreaction*:** Discussed with psychotherapy.

(o) ***Contracts*:** It is often the case that the reinforcing consequences of a patient's behaviour are under the control of another person.

Yoga and Meditation

The term "Yoga" is derived from the root word "Yuj," meaning union. The worldly approach describes it as the union between the mind and the body. The spiritual approach, on the other hand, regards it as the union between the individual self and the cosmic self. The philosophical explanation was cited in the *Vedas* and *Upanishads,* which conceived of the world, the *Atman,* as a conscious spiritual principle permeating all things.

The systematisation of this knowledge and practice was formulated by the great Indian seer Patanjali (200 A.D.) in his *"Yoga Sutra"*.

Based on this broad conceptual framework, many yogic procedures have been developed in India. The basic components of all different yogic schools are: (1) yogic teachings (*Yama* and *Niyama*), (2) *Kriyas,* yogic postures, *Bhanda* and *Mudra (Asana),* (3) breathing exercises *(Pranayama),* and (4) meditations *(Dhayana, Dharana* and *Samadhi*). Earlier, the practice of yoga was to attain spiritual and mystic goals. Later, yoga was practiced as a psycho-physiological technique for voluntary control and for integrating the body and the mind.

Recently, yoga has been recognised worldwide as a treatment procedure. Yoga therapy has been tried in psychiatric as well as psychosomatic disorders such as anxiety neurosis, psychogenic headache, depression, hypertension, bronchial asthma, diabetes etc.

Meditation and Psychotherapy

Appraising Indian philosophical systems in general and the yoga system in particular, Wolberg observed that yogic approaches are a combination of supportive and educational modalities. In contrast, transpersonal theorists contend that meditation may provide inner calm, loving kindness towards oneself and others, access to previously unconscious material, transformative insight into emotional conflicts, and changes in the experience of personal identity.

Yoga and Behavioural Therapy

In the West, yoga is seen from a behaviourist viewpoint, and most of the studies compare yoga with behavioural techniques. Certain behavioural methods such as progressive relaxation, autogenic training, and reciprocal inhibition are considered to be similar to *Shavasana, Pranayama* and *Samyama* meditation. The procedures in yoga such as *Yama* and *Niayama* are like the stimulus control and response control aspects of the environment, which are used in self-control procedures of behavioural therapy.

Psychophysiological Correlates of Meditation and Yoga

Later researchers hypothesised that meditation produces a "hypometabolic state", with such changes as decreased oxygen consumption, decreased blood pressure, increased skin resistance, and increased coherence of electrical activity of the brain.

Psychophysiological Correlates of Yoga

Studies on yoga practice, which included *Kriyas, Asanas, Pranayama,* and *Meditation,* revealed significant alterations in physiological functions.

Studies of full yogic practices showed an increase in muscular, cardiovascular and respiratory efficiency and decreased blood pressure and heart rate.

Several studies of yoga have found an increase in plasma proteins, corticosteroid metabolites, plasma cholinesterase, and volume, acidity, and total solids in urine. A decreasing trend in blood sugar and specific gravity of urine was observed with the practice of yogic exercises and pranayama.

Psychological Correlates of Meditation and Yoga

The major findings were an increase in self-actualisation, interpersonal relations, locus of control, relaxed nature, time, competency, inner directedness, and self-worth and better emotional and home adjustment. Cognitive functions like immediate memory, visuomotor coordination, and visual and auditory reaction time were improved with the yoga practice.

Effect of Meditation and Yoga in Neurotic and Psychosomatic Illnesses

Significant improvement has been reported in patients with anxiety reaction, neurotic depression, obsessive compulsive disorder, or involutional depression whereas less effective results are seen with schizophrenia and personality disorders.

Recent reports on bronchial asthma showed such promising results as an increase in vital capacity, peak expiratory flow, and a decrease in medication requirement.

Since yoga is grounded in a cultural-religious milieu, it involves not only postures, breathing exercise, or meditation but also yogic diet, faith, and a way of spiritual living. As opposed to drugs, these have the advantage of being inexpensive, easily accessible, devoid of side effects, and useful in preventive and promotive roles.

B. BIOFEEDBACK AND BEHAVIOURAL MEDICINE

Behavioural medicine is a term introduced by *Birk*s in 1973.

Definition

Behavioural medicine is the *interdisciplinary* field concerned with the development and *integration* of behavioural and biomedical science knowledge techniques relevant to health and illness and the application of this knowledge and these techniques to prevention, diagnosis, treatment and rehabilitation.

Theory

Feedback from the environment about the consequences of one's acts provides the rewards and punishments that are an important part of learning.

Applications of Psychophysiologic Problems for which a Single Biofeedback Modality is Useful

(a) *Thermal Feedback (applied to circulatory and sympathetic nervous system disorders)*
- — Migraine Headache
- — Raynaud's Disease
- — Buerger's Disease

(b) *EEG Feedback (applied to central nervous system problems)*
- — Epilepsy (GME)
- — Creativity Enhancement
- — Narcolepsy

(c) *EMG Feedback (applied to disorders involving muscle function)*
- — Tension Headache
- — Neuromuscular Re-education
- — Cerebral Palsy (symptomatic relief)
- — Bruxism

— Subvocalisation
— Blepharospasm
— Bell's Palsy
— Torticollis
— Spasticity (symptomatic relief)
— Tics

C. SOCIAL SKILLS TRAINING

It is a type of behaviour therapy used in persons who have marked deficits in social skills (e.g. schizophrenics) or who have developed adequate social skills appear to lose them (e.g. a psychiatric disorder) or fail to employ them aptly to achieve their social goals (e.g. inadequate personality).

Terminology

"Assertiveness" and "Assertive training". Wolpe (1958) recognised that many patients with social anxieties are unassertive particularly in the sense of not "standing up for their rights".

Wolpe proposed the terms "hostile" assertiveness (persons who fail to correct perceived interpersonal injustices and wrongs) and "commendatory" assertiveness.

Indications

— Schizophrenia
— Depression (Unipolar or bipolar disorder)
— Anxiety disorders (e.g. Anxiety neurosis, phobias, obsessive compulsive disorder, post traumatic stress disorder etc)
— Alcoholism
— Mental retardation or emotionally disturbed children.
— Aggressive behaviour problem.
— Inadequate Personality.
— Non-clinical applications e.g. in marital relationships, occupational problems, other intense dyadic relationships.

Treatment Techniques

(i) Instruction and Coaching
(ii) Modelling
(iii) Behavioural Rehearsal (guided practice, role playing)
(iv) Feedback
(v) Social Reinforcement
(vi) Homework Assignments